Proceedings of the Linnean Society of New South Wales

Volume 138

Proceedings of the Linnean Society of New South Wales
Volume 138

ISBN/EAN: 9783744657273

Printed in Europe, USA, Canada, Australia, Japan

Cover: Foto ©berggeist007 / pixelio.de

More available books at **www.hansebooks.com**

THE LINNEAN SOCIETY OF
NEW SOUTH WALES
ISSN 1839-7263

Founded 1874
Incorporated 1884

The society exists to promote the cultivation and study of the science of natural history in all branches. The Society awards research grants each year in the fields of Life Sciences (the Joyce Vickery fund) and Earth Sciences (the Betty Mayne fund), offers annually a Linnean Macleay Fellowship for research, and publishes the *Proceedings*. It holds field excursions and scientific meetings including the biennial Sir William Macleay Memorial Lecture delivered by a person eminent in some branch of natural science.

Membership enquiries should be addressed in the first instance to the Secretary. Candidates for election to the Society must be recommended by two members. The present annual membership fee is $45 per annum.

Papers are published at http://escholarship.library.usyd.edu.au/journals/index.php/LIN and access is free of charge. All papers published in a calendar year comprise a volume. Annual volumes are available to any institution on CD free of charge. Please notify the Secretary to receive one. "Print on demand" hardcopies are available from eScholarship.

Back issues from Volume 1 are available free of charge at www.biodiversitylibrary.org/title/6525.

The postal address of the Society is P.O. Box 82, Kingsford, N.S.W. 2032, Australia.
Telephone and Fax +61 2 9662 6196.
Email: linnsoc@iinet.net.au
Home page: www.linneansocietynsw.org.au/

Cover motif: Tenison Woods, President of the Linnean Society of New South Wales, 1879-1881.

PROCEEDINGS
of the
LINNEAN
SOCIETY

of
NEW SOUTH WALES

For information about the Linnean Society of New South Wales. its publications and
activities, see the Society's homepage
www.linneansocietynsw.org.au

VOLUME 138
December 2016

An Ecological History of the Koala *Phascolarctos cinereus* in Coffs Harbour and its Environs, on the Mid-north Coast of New South Wales, c1861-2000

Daniel Lunney[1], Antares Wells[2] and Indrie Miller[2]

[1]Office of Environment and Heritage NSW, PO Box 1967, Hurstville NSW 2220, and School of Biological Sciences, University of Sydney, NSW 2006 (dan.lunney@environment.nsw.gov.au)
[2]Office of Environment and Heritage NSW, PO Box 1967, Hurstville NSW 2220

Published on 8 January 2016 at http://escholarship.library.usyd.edu.au/journals/index.php/LIN

Lunney, D., Wells, A. and Miller, I. (2016). An ecological history of the Koala *Phascolarctos cinereus* in Coffs Harbour and its environs, on the mid-north coast of New South Wales, c1861-2000. *Proceedings of the Linnean Society of New South Wales* 138, 1-48.

This paper focuses on changes to the Koala population of the Coffs Harbour Local Government Area, on the mid-north coast of New South Wales, from European settlement to 2000. The primary method used was media analysis, complemented by local histories, reports and annual reviews of fur/skin brokers, historical photographs, and oral histories. Cedar-cutters worked their way up the Orara River in the 1870s, paving the way for selection, and the first wave of European settlers arrived in the early 1880s. Much of the initial development arose from logging. The trade in marsupial skins and furs did not constitute a significant threat to the Koala population of Coffs Harbour in the late nineteenth and early twentieth centuries. The extent of the vegetation clearing by the early 1900s is apparent in photographs. Consistent with the probable presence of Koalas in the Coffs Harbour town centre in the early 1900s, available evidence for the period 1920-1950s strongly suggests that Koalas remained present in the town centre and surrounding area. Large-scale development began in the early 1960s. Comparing aerial photographs allows us to discern the speed of change from a largely rural landscape in 1964 to one that is predominantly urbanised by 2009. The 1999 Comprehensive Koala Plan of Management for Coffs Harbour City Council, drawing on the 1990 Community Survey of Koalas in Coffs Harbour, detailed specific examples of habitat fragmentation through development. Local media coverage offered a wealth of information on the persistence, and rapid eradication, of Koala habitat over the 1970s-2000, in addition to the level of community interest in the issue. Taken collectively, the evidence allows us to draw two main conclusions: that the Koala population of Coffs Harbour was widespread but never abundant, and that habitat loss has been relentless since European settlement. The transformation of a rural-forest to a largely urban landscape, particularly in the south-east of the Local Government Area, over the past four decades is the most recent stage in the incremental loss of habitat since European settlement. Consequently, the conclusion can be drawn that the Koala population had been reduced from its pre-European size by 2000. Concurrent research on the Coffs Harbour Koala population showed that it declined during the 1980s, but was relatively stable and endured over the period 1990-2011. These findings point to the necessity of employing historical analysis to interpret change in Koala populations in Coffs Harbour to complement current assessments of population status.

Manuscript received 24 October 2014, accepted for publication 21 October 2015.

Keywords: Bellingen, Coffs Harbour, *The Coffs Harbour Advocate*, ecological history, fur trade, Koala, Orara, *Phascolarctos cinereus*, media analysis, native bear, timber industry, vegetation clearing.

INTRODUCTION

This paper aims to develop an ecological history of the Koala *Phascolarctos cinereus* in Coffs Harbour on the mid-north coast of New South Wales, focusing primarily on changes to its population profile since European settlement in the region. It forms part of a series of papers that aims to track the population in order to interpret its current ecological status. The first comprehensive, Shire-wide Koala Plan of Management in NSW was prepared for Coffs Harbour City Council in 1999 (Lunney et al. 1999a, 2000, 2002) and adopted in State Parliament in 2000. In evaluating this Plan, we considered it essential to analyse not only the recent profile of the Koala population, but also the pattern of long-term change. In order to fully understand the long-term trend as well as to interpret the current status of the Coffs Harbour Koala population, we must adopt both an ecological and an historical approach that goes beyond three Koala generations, which is 20 years (Australian Government Department of the Environment 2011). The historical enquiry undertaken in this paper provides the context within which ecological interpretations of the long-term changes in and current status of the Koala population of Coffs Harbour can be viewed.

In view of a number of methodological challenges and evidentiary deficiencies that emerged in the research process, this paper makes no claim to being exhaustive. Rather, it proposes a thesis of the general pattern of historical change with regard to the Koala population of Coffs Harbour, to complement the intense ecological work currently being undertaken (Lunney et al. 2015). In so doing, it corresponds to an historical approach which aims to track and interpret the long-term pattern of animal population changes in relation to the pattern of human settlement over longer time frames than those generally regarded as long-term in ecological research, i.e. 10 years or more. This framework is far from definitive and one of the objectives of our work is that it may inspire other scholars to refine it in their efforts to trace the elusive changes of animal populations across historical time, with an eye to the interaction between them and human settlement.

Concurrent ecological research has identified that the Koala population of Coffs Harbour has persisted over the period 1990-2011 both in terms of distribution and activity levels, and that it is, surprisingly, relatively stable (Lunney et al. 2015). This follows on from a population decline in the 1980s. These conclusions arose from two independent survey methods (community survey and field survey). There are four possible explanations for the Koala population's stability from 1990-2011: that recent conservation efforts and planning regulations have been effective; that surviving adults are persisting in existing home ranges in remnant habitat; and that the broader Coffs Harbour population is operating as a "source and sink" metapopulation, with nearby higher density populations (such as Bongil Bongil National Park) providing a source of immigrant Koalas; and/or that the standard survey methods employed are not sufficiently sensitive to detect small population changes (Lunney et al. 2015). The present paper is intended to deepen our understanding of the long-term profile of the Coffs Harbour Koala population and extend our focus beyond the last three decades.

Looking at population trends over long time periods provides a deeper understanding of possible drivers of population change, thereby allowing better future management of the remaining population.

METHODOLOGICAL NOTES

Ecological history is a rapidly growing field attracting considerable international attention. Drawing on existing fields such as environmental history and historical geography, ecological history has been recognised as crucial to developing ecological restoration programs and conservation strategies (Foster 2000; Donlan and Martin 2004; Jackson and Hobbs 2009). As a discipline it requires both ecological and historical understanding, utilising the analytical tools and approaches of both ecology and history to shed light on the relationships between humans and the natural environment. Many works in the field adopt a grand-scale approach, examining ecological changes which have taken place over millennia in whole regions (e.g. Vermeij 1987; Flannery 2001; Grove and Rackham 2001). For more localised studies, however, an approach on a smaller scale is equally valuable in capturing the ecological specificities and changes of a given area.

Ecological histories of Australian fauna are rare. Of those that exist, we can discern a number of general approaches and research foci. Studies that examine specific species from a management perspective that takes historical data into account are rare (for example, see Menkhorst 2008). Others examine the impact of a specific exploitative activity on a species, such as the trade in seal skins in south-eastern Australia (Ling 1999) and marsupial furs (Koalas and Brushtail Possums *Trichosurus vulpecula*) in Queensland (Hrdina and Gordon 2004; Gordon and Hrdina 2005). One short essay (Parris 1948) attempts to track changes in Koala abundance on the Goulburn River, Victoria, using historical sources, but it is neither comprehensive in its research nor rigorous in methods. Multiple studies have utilised historical data to assess decline in species distribution (Lunney et al. 1997; Lunney 2001; Gordon et al. 2006). The majority of studies adopt a state-wide scale, aiming to identify general patterns of change and/or infer local population changes from this picture.

While state-wide analyses allow us to contextualise regional changes within broader historical patterns, these broader patterns do not always align neatly with the patterns of specific districts within the state. Following the first, comprehensive state-wide survey of Koalas in New South Wales (NSW)

2

Proc. Linn. Soc. N.S.W., 138, 2016

in 1986-87, Reed and Lunney (1990) concluded that habitat loss was the most decisive factor in the decline of the Koala in NSW. However, as the settlement of NSW varied across localities due to geographical specificities, so did the severity and timing of the impact on fauna and the natural environment. Research on the Koala populations of Campbelltown (Lunney et al. 2010), Port Stephens (Knott et al. 1998) and Bega (Lunney and Leary 1988) indicates that, while Koala populations responded similarly to settlement, the impact differed widely among these districts. Additionally, we must also consider other factors, such as the varying impact of the fur trade, the varying densities of the initial Koala populations, and the extent of Koala occurrence across a given geographic range.

An appreciation of these variables is critical in developing a comprehensive understanding of the management and restoration challenges that face a species. Long-term studies of fauna undertaken prior to 1960 are rare worldwide and, as one recent paper notes, this is particularly the case with regard to quantitative studies in historical ecology more broadly (Zu Ermgassen et al. 2012). Consequently, predictions of a bleak future for a species are generally only based on the analysis of the last 30-50 years of that species' occurrence. The alternative to these alarmist (and, in some cases, fatalistic) assessments is complacency: here, short-term data facilitate the conclusion that continued management and/or restoration programs are unnecessary. In order to avoid potential oversights, we must develop an historical understanding that takes into account the patterns of recent decades, but that is not restricted to them.

Recent advances in our collective knowledge of Koala ecology, and the threats Koalas face, have sharpened our focus as to what environmental and ecological attributes are likely to have influenced the changes to Koala populations. This allows us to be more inclusive in our research, by enabling us to identify and examine factors previously overlooked in considerations of long-term population change, such as the impact of the fur trade. It also allows us to be more precise in our analysis and interpretation of historical sources, and critical of their relative significance in the context of the Koala population. In addition, the Koala has an attribute which makes it a near perfect animal to study historically. There is only one species of Koala, so there is no confusion about what species is mentioned in various historical documents of a non-scientific nature, such as newspaper reports. The Koala is large and slow-moving, so when it is seen, it can be readily identified. As they are obligate tree-dwellers, and as their forest habitat is logged or cleared for housing, their populations can be tracked by looking at changes to the habitat on which they depend. This makes the Koala an ideal species to look at through indirect evidence from an ecological viewpoint.

However, while the Koala is distinctive, the historical evidence allows limited scope for interpreting change in population size over time. Numerical data, such as might be expected in a scientific study, were not available prior to 1990 when a systematic survey was undertaken across the Coffs Harbour Local Government Area (LGA) (Lunney et al. 1999a). As a result, we can expect only large shifts in numbers to be registered in historical records. As will become apparent, the recorded 'changes' are imprecise, unsystematic, and generally refer to perceived numbers as opposed to distribution. The 1990 study was also the first to determine the range and habitat preferences of the Coffs Harbour Koala population (Lunney et al. 1999a). Prior to 1990, historical sources identify specific locations but do not offer a systematic assessment of distribution across the LGA. Consequently, when we utilise the term "population" in this paper, we refer to numerical size or abundance rather than shifts in distribution, unless otherwise indicated.

In the context of Coffs Harbour, the scarcity of long-term data lends living memory a particular significance. As oral histories of older residents were not conducted until the mid-1990s, the experiences related in their accounts are weighted towards the latter half of the twentieth century. Nevertheless, they comprise an important point of comparison with existing sources from the period. In addition, the 1990 Koala Survey (Lunney et al. 1999a, 2000), conducted by the National Parks and Wildlife Service as a response to recommendations made at the 1988 Koala Summit (Lunney et al. 1990), provides us with a crucial source of perceptions data (i.e. memory and perception of the past and current presence of Koalas locally, and of the issues facing Koalas) for the Coffs Harbour area. As the diversity of most species leads to popular confusion, perceptions data can generally only be utilised for a single species, and usually iconic species at that. Due to its distinctiveness, the Koala is one of the few Australian animals that can be reliably identified by non-specialists. This paper draws heavily from the respondents' comments, particularly those of long-term residents, in order to substantiate its broader analysis of the extent of Koala occurrence in the area over time. Though these comments rest entirely on the respondents' memories, we maintain that, despite the likelihood of potential errors in individual accounts, these comments support our general thesis of Koala occurrence when taken in the aggregate. These perceptions data are intended to complement information gleaned from other sources, primarily newspaper reports, historical photographs, and local histories of the area.

The primary method utilised in this study is media analysis. For the purposes of this study, this method involves the comprehensive reading of newspapers from the period in order to gauge the changing profile of the Koala population in the Coffs Harbour area. It also requires us to pay attention to the ratio of information about Koalas in comparison with other animals. In this regard, it is instructive to note the criteria applied to the media coverage of animals, which remains relatively consistent throughout the period examined in this study: generally, animals do not warrant coverage unless they are considered pests (and thereby threaten the stability of human practices), they carry a 'scare value' (and are thereby perceived to threaten human life), or they are commercially important. With this in mind, the relative silence about Koalas in the local print media of the Coffs Harbour area is itself historically interesting, for it suggests that residents did not view Koalas as pests, unlike paddymelons (small members of the kangaroo family) and flying-foxes, nor were they an important trade item in the area. It also suggests that residents were not particularly interested in their welfare until the 'conservation turn' of the late 1960s.

Media analysis also requires us to pay attention to coverage of the practices that affect Koala habitat, such as vegetation clearing, ringbarking, and the fur trade. Shipping reports for the area, and for the steamers which utilised the Coffs Harbour port, could not be located and have presumably been destroyed. As a result, we have had to rely on the reports printed in local newspapers for information regarding the exports that passed through the Coffs Harbour port. These reports take the form of summaries and are intended to publicise the 'going rates' of key exports. However, as Coffs Harbour's local paper, the *The Coffs Harbour Advocate*, began in 1907, the reports up until this point have been drawn from two regional newspapers. These papers are *The Clarence and Richmond Examiner and New England Advertiser* (published 1859-1889) and *The Clarence and Richmond Examiner* (published 1889-1915) [*hereafter, in-text citations of these papers will take the forms of CRENEA and CRE, respectively*]. Both were published in Grafton, a town on the Clarence River, north of Coffs Harbour. As the distribution of these papers stretched from the Tweed, in the north of the State, to Bellingen, immediately to the south of Coffs Harbour, and included townships as far west

as Tamworth and Armidale, it is difficult to discern precisely from where the skins and furs listed in the shipping reports originated. It is also not known exactly how many skins comprised a bale. As such, these reports give us only a partial indication of the extent of the fur trade in the Coffs Harbour area. Furthermore, as the *The Coffs Harbour Advocate* [*hereafter, in-text citations of The Coffs Harbour Advocate will take the form of CHA*] was issued daily, we have adopted a 'sampling' approach in view of time constraints. By reading the issues for the first year of each decade (e.g. 1910, 1920, 1930), we aim to identify general patterns of change.[1]

In view of the limitations of media analysis, we have consulted local histories of Coffs Harbour, in addition to the nearby settlements of Bellingen, Raleigh, and Urunga; histories of the fur trade in Australia; historical photographs; and accounts of shipping on the mid-North Coast. We have also searched other local and regional newspapers for articles on the fur trade, the development of Coffs Harbour, and reports on trips to the mid-North Coast taken by commissioned explorers. In order to situate Coffs Harbour in the broader fur/skin market, we consulted the following annual reviews of fur/skin brokers: Goldsbrough, Mort and Co.; Dalgety and Co.; Winchcombe, Carson and Co.; and Bridge and Co. Regrettably, the archives of the Coffs Harbour Historical Society were inaccessible during the research stage of this paper due to extensive flooding, which forced the Society to store its archival material in shipping containers for an indefinite period.

EARLY HISTORY

Historical sources concerning the fauna of the Coffs Harbour region prior to European settlement are scarce. While scholars of Aboriginal history have identified that the dialects of the Gumbaynggir nation contain multiple words for 'native bear' (Ryan 1988: 23-24), this tells us little about the precise distribution of the Koala in the region, for the speakers of these dialects are located not only in the Coffs Harbour area but also as far as Grafton and Nambucca Heads (a township south of Coffs Harbour) and Bellingen. On a slightly smaller scale, the "tribal territory", or the area of land recognised as the "particular preserve" of the Gumbaynggir nation, has been estimated at 6,000 square kilometres (Ryan 1988: 56, c.f. Tindale 1940). The Coffs Harbour town area tribe was known as 'Womboyneralah', or "where the kangaroos camped" (England 1976: 46). Among the words for 'native bear' in the Gumbaynggir lands are 'Toon-gari', which is a word specific to the inhabitants of the Orara (Rudder 1899), and 'Yarrahapinni', specific to the Macleay and meaning 'native bear rolling down the hill' (Tyrrell 1953).

With regard to the cultural practices and beliefs of the Gumbaynggir peoples, multiple sources indicate that they utilised Koala skins to make rugs (McFarlane 1934b; Yeates 1990: 11; Thomas 2013). This is consistent with an 1880 newspaper report on the Australian fur trade, which notes that the best Koala fur was put to this use across Australia (*The Argus* 1880). Yeates (1990: 10) notes that the tribes of the mid-North Coast were versed in a particular method of tree-climbing, which allowed them to obtain "honey, opossums and koalas". Koalas are prominent in mythologies relating to the North Coast, such as the legend of the "great bear" of Mount Yarrahapinni (Ryan 1988: 114, 125) and the legend of Ulitarra, which is but one of many legends connecting the Koala with water and, at times, salvation from danger (Ryan 1964). Furthermore, it was a totemic feature for at least two tribes in the area (Ryan 1988: 51). It is difficult to discern if the Gumbaynggir peoples commonly ate the Koala and used its skin: while we can assume that it was widely hunted (McFarlane 1934a), it is conspicuously absent from the explorer Clement Hodgkinson's account of the tribes he encountered along the Bellinger River, in which he noted animals they consumed (Hodgkinson 1845: 43, 45, 58). In an account of the food regulations of the Gumbaynggir, Ryan (1988: 53) notes that, immediately after a young male was admitted to the status of a tribesman, he was "often forbidden to eat the male of the native-bear, kangaroo, opossum, or short-nosed bandicoot". It is unclear, however, whether this custom was consistent across tribes in the Gumbaynggir lands.

European settlers reached Coffs Harbour relatively late in comparison to other areas of the north coast of New South Wales. The movements of the settlers depended upon their ability to safely transport themselves, their cargo and exports via sail (and later steam) vessels. The absence of a river connecting Coffs Harbour to the sea meant that it was left out of the initial phase of settlement of the North Coast in

the 1830s and 1840s (England 1976: 6). This phase of settlement was confined to Bellingen on the Bellinger River and Grafton on the Clarence River. Until 1830, free settlement to the north of Port Macquarie was prohibited, leaving Bellingen undisturbed. When this ban was lifted, settlers faced the challenge of crossing a "hazardous" bar at the mouth of the Bellinger River and "an almost impenetrable forest" (Pegum and Pegum 2010: 16). In contrast to the flat, fertile country of the Clarence and Bellinger Rivers, Coffs Harbour's considerable elevation (see Fig. 1) meant that it was largely inaccessible, and poor land for harvesting crops. Geography was thus a primary determinant of the pattern of the initial settlement of the Coffs Harbour area and of the mid-North Coast in general.

The first Europeans in the Coffs Harbour area have been variously reported as escaped convicts "taking refuge" on Muttonbird Island (Rodwell 2011: 27), and two sailors who wandered away from their ship in 1837 and followed the Orara River (Secomb 1986: 4). In 1840, a stockman named William Miles, employed by a Macleay grazier, headed north with the intention of identifying new rivers alongside which cedar grew in abundance. His glowing reports of the cedar near the Bellinger River persuaded Clement Hodgkinson, then the Government Surveyor of the Macleay District, to explore the area for himself (Hobson 1978: 4). Hodgkinson undertook two expeditions to the area in 1841-42, and his account of what he observed later formed Part 1 of his book *Australia from Port Macquarie to Moreton Bay* (Hodgkinson 1845). Fauna is largely peripheral to his account, which focuses on the vegetation and geology of the Bellingen area and, to a lesser extent, the local Aboriginal tribes he encountered along the route. While he is highly attentive to the appearance, abundance, and utility of the natural resources he observes, animals do not receive the same degree of analytical interest: only once does an animal – a kangaroo – appear in the narrative because it constitutes, *in itself*, an interesting feature of the landscape (Hodgkinson 1845: 47). With regard to other passages in which animals are mentioned, the

Fig. 1 (following page). General goods, timber, and cedar transportation routes in the local area, showing elevation and the place names mentioned in the text. These transportation routes were established by the late 1880s. This digital elevation model map shows that the Clarence Valley and the Clarence River were located on a large expanse of low-lying land. Similarly, the Bellinger Valley to the south of Coffs Harbour is on low-lying land. The high elevation of Coffs Harbour, other than the coastal strip, is also evident. The trade routes of the 19th and early 20th centuries are shown, as are key towns in this early trading settlement. This map places Coffs Harbour in its regional context, and allows us to understand why Coffs Harbour, being hilly with no river, was settled much later than the Bellinger and Clarence Valleys. This map is based on information derived from historical sources.

majority are in connection with commentary on the hunting practices of the Aborigines who accompany Hodgkinson on his expedition (see e.g. Hodgkinson 1845: 45, 58, also Part V). Interestingly, the animals which the Aborigines of the Bellinger area hunt and consume include "a kangaroo", "a carpet-serpent", "pademella", a "brush-kangaroo", "an opossum and a large dew-lizard", but not the Koala (Hodgkinson 1845: 28, 30, 33, 43, 45).

IMMEDIATE PERIOD OF EUROPEAN SETTLEMENT: 1870–C.1890

In 1847, shipbuilder John Korff, on his way to the Bellinger River, sought refuge from a gale in a port which he named Korff's Harbour. Although he reported his discovery after returning to Sydney, European settlement of the area did not commence until the mid-1860s. Following the passage of the Robertson Lands Act in 1861, the NSW Government reserved the land adjacent to the Harbour (NSW Government Gazette 1861), evidently recognising its potential as a port. The Bellinger Valley was opened to selection in 1863. While an early pioneer arrived in 1865 to draw cedar (Yeates 1990: 20), the first wave of settlers did not arrive in Coffs Harbour until the early 1880s. Cedar-cutters gradually worked their way up the Orara River in the 1870s, paving the way for selection. The discovery of gold in the Orara Valley in 1881 hastened the arrival of newcomers to the area (Yeates 1990: 23). By 1890, a small but thriving community of selectors, sawmillers and teamsters had developed in Coffs Harbour (Bacon 1926: 96).

Local histories of Coffs Harbour show that, in the early period of settlement, the development of crops was modest (Yeates 1990; England 1976). As a result, much of the initial development of the Coffs Harbour hinterland arose from extractive industries, primarily logging. The timber industry experienced a rapid boom in the early 1880s. By March 1883, the newly-appointed Inspector of Forests noted that the number of sawmills in the area was steadily increasing, with five more about to be established (Secomb 1986: 8). Furthermore, following an inspection of the Orara reserve, he recommended that if the reserve was not to be retained, a "corresponding area" containing a "similar description of brush forest" should be reserved in its place, "otherwise many of the bush timbers are likely soon to become extinct" (Duff 1883, c.f. Secomb 1986: 8). Three years later, merely a few weeks before the Parishes of Coff and Wongawonga were opened to selection, *The Clarence and Richmond Examiner and New England Advertiser* obtained a description of the uncultivated land:

The soil is rich alluvial in the flats, and fair arable land on the lower ridges; the back ridges are generally steep, broken and stony. The whole of the land with the exception of the caps of one or two ridges is covered with scrub dense in the flats and dense to light on the back ridges. The timber is plentiful and good, consisting of flooded and red gum, box, bloodwood, oak, tallowwood, blackbutt, and several varieties of scrub woods (softwoods) [...]. (*CRENEA* 1886)

After the Parish of Coff was opened to selection in July 1886, clearing became more extensive. Settlers often employed local Aboriginal people to assist in clearing the tangled undergrowth, with the animals brought down in the 'drive' – primarily hundreds of flying-foxes – serving as recompense (England 1976: 17). The felled timber rapidly became the area's foremost export, shipped via the port of Coffs Harbour to Sydney (Richards 1996: 78-81). The local and statewide timber transportation routes are shown in Figs 1-2.

While the commercial significance of timber for the early Coffs Harbour community is indisputable, it is more difficult to ascertain the importance of the trade in marsupial skins and furs for the settlers. England (1976: 18) notes that the settlers' guns "were seldom idle", listing the Koala alongside wallabies, possums and kangaroos as an animal killed for its skin in Coffs Harbour, but provides no source for his claim. As Coffs Harbour lacked a local paper at this time, we must turn to regional papers for insight into the magnitude of the fur trade in the area. Commercial and shipping reports published in *The Clarence and Richmond Examiner and New England Advertiser* and *The Clarence and Richmond Examiner* show that there was an active fur trade in the broader region. Due to the wide distribution of these papers, it is difficult to determine the specific import of their commentary for Coffs Harbour. At the very least, the papers publicised the prices for skins, thereby informing settlers of their fluctuating value and allowing them to develop reasonable estimates of the returns they could expect from a hunt. With this function in mind, it is important to note that the commercial reports evince a low 'going rate' for Koala skins in comparison to the skins of other marsupials. In one report in late 1889, a fur/skin broker lists the "Bear" at "1d to 3.5", while a large Grey kangaroo fetches "80d to 95d" and the Swamp wallaby "7d to 19d" (*CRE* 1889j). Kangaroo fur was undoubtedly the most popular and consistently commanded the

Proc. Linn. Soc. N.S.W., 138, 2016

7

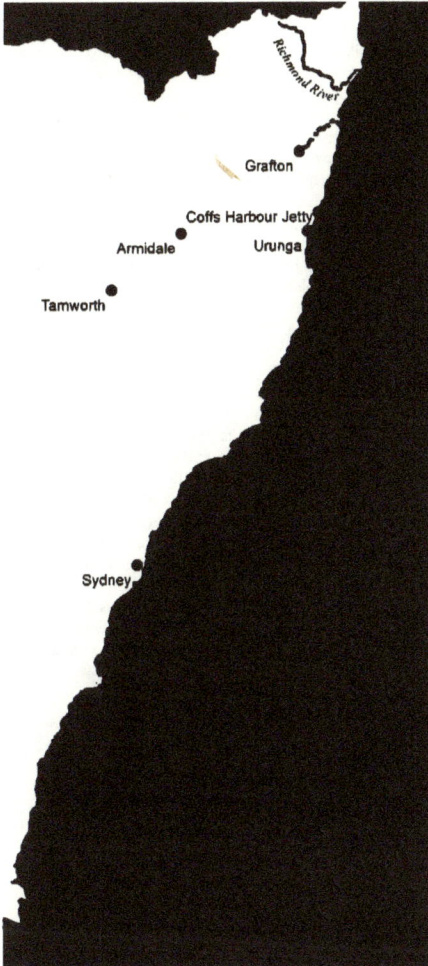

Fig. 2. General goods, timber, and cedar transportation routes between the North Coast townships under examination and Sydney. These routes were established by the late 1880s. Cedar logged in the Bellinger Valley was transported to Coffs Harbour. Urunga is at the mouth of the Bellinger River. Grafton was an important point of mid-north coast settlement. Grafton and Coffs Harbour were linked by trade. This map is based on information derived from historical sources.

highest prices (*CRE* 1889a,c,e,f,g,h,i,k). Indeed, the demand for kangaroo was so great that some worried it could become extinct in the area and recommended a closed season (*CRE* 1889b,d). Opossum fur also sold well, though prices were subject to its quality (*CRE* 1889g,h,j).

Though price was not the sole incentive for hunting a particular animal (Fuchs 1957), the low rate for Koala skins can be considered particularly dissuasive when viewed in conjunction with the relative inaccessibility of Koalas. This is shown in an article originally printed in the *Tenterfield Record* and republished in *The Clarence and Richmond Examiner and New England Advertiser* (1889) under the headline "The Skin Trade". The correspondent, based in Tabulam on the lower Clarence, writes:

> A small party of men from Tenterfield arrived here last Thursday for the purpose of procuring bear (and I believe opossum) skins for one of your storekeepers. They lost no time in commencing their operations, and pitched their camp on the Clarence, at its confluence with the Timbarra, whence they despatched in every direction a number of blackfellows whom they had engaged upon their arrival here. I believe they have not met with any extraordinary amount of success in their undertaking of nabbing the agile koala. Prices do not range very high for the skin of this festive and beautiful creature, and it would require at least 250 or 300 a week to liquidate current expenses; probably more would be necessary.

Contrary to Marshall's (1966: 26) characterisation of the Koala as a "sitting duck", the article shows that hunting Koalas required great skill. It also shows that its commercial returns were disproportionate to the effort and resources expended by the hunters, which presumably included a payment, likely of goods, to the Aboriginal people who assisted them. Furthermore, we may surmise that there were enough Koalas readily available across the state to maintain non-competitive prices, in comparison to kangaroos and wallabies, which were the objects of hunting pressure from the early 1870s in NSW, particularly in the south (Lunney et al. 1997). Here, it is interesting to note that international accounts of the fur trade emphasise that the fur of the Koala is "cheap" (Poland 1892: 365) and "not as important commercially as the Common Phalanger [opossum]", though useful "where a durable, reasonable priced fur is desired" (Petersen 1914: 263).

It is extremely difficult to develop an historical baseline for the Koala population of Coffs Harbour at European settlement from the extant sources. Indeed, to the best of the authors' knowledge, there is only one primary source for this period that mentions the Koala specifically *and* is connected to the emerging township of Coffs Harbour. It is a short advertisement placed in *The Clarence and Richmond Examiner* by Hermann Rieck, a selector who settled at Coffs Creek in 1881:

> Young native bear for sale. Able to keep himself on gum leaves; very tame, and easy to be transported in a bag on saddle. Will sleep for days without any noise or disturbance. H. Rieck, Coffs Harbour. (*CRE* 1886a)

The advertisement was published in two consecutive issues (*CRE* 1886a,b) after which Rieck's Koala presumably found a home. It shows that, at this time, Koalas were rare enough – or, alternatively, undesirable enough – in the area to warrant a public attempt to sell them. It is possible that Rieck appealed

to readers of the regional paper because he could not sell the Koala in his immediate vicinity: why, after all, would he pay to place an advertisement in a paper when he could inform his friends and neighbours free of charge? Having been one of the first settlers in Coffs Harbour, he was well known in the area. Regardless, judging by his description of the Koala, it appears to not have occurred to Rieck that he, or others, could sell it on the fur/skin market. Instead of the quality of its fur, its domicile nature takes precedence in his description. Most interestingly, photographs of Rieck's homestead and banana plantation, taken in the early 1890s, show that both were surrounded by Koala habitat (Figs 3 and 4).

In view of the scarcity of records concerning Koalas for this early period of Coffs Harbour's history, we can draw few definitive conclusions. We can be fairly certain, however, that the rapid clearing spurred on by the growing timber industry led to the beginning of the fragmentation of Koala habitat in the Coffs Harbour area. Secondly, we may surmise that the trade in marsupial skins and furs, which accelerated in the 1880s and was active in the broader region, produced flow-on effects for the Koala population, though we cannot specify the nature or extent of these effects. While the absence of specific records regarding skins does not permit the conclusion that the trade was minimal in Coffs Harbour, Rieck's advertisement allows us to assume that, in 1886, the trade was not so prominent in the area as to have precluded an attempt to sell a Koala

Fig. 3. Hermann Rieck's homestead, Korora, c1890s. Reproduced courtesy of the Coffs Harbour Library and Coffs Harbour Regional Museum. Accession no. 07-4760.

Proc. Linn. Soc. N.S.W., 138, 2016

9

Fig. 4. Hermann and Fanny Rieck on their banana plantation, c1892. Reproduced courtesy of the Coffs Harbour Library and Coffs Harbour Regional Museum. Accession no. 07-2421.

by other means. Drawing on the existing records, we may conclude that Koalas were present in Coffs Harbour in the early period of European settlement, but arguably not in high numbers.

GROWTH AND EXPANSION: 1890-1920

Coffs Harbour's timber industry underwent rapid growth in the 1890s and early 1900s. After the construction of the Coffs Harbour jetty was completed in 1892, facilitating the export of hardwoods, timber-cutters "flocked" to the district (England 1976: 16). Many settled close to the centre of town, building homes of flooded gum and beech, and by late 1892 all of the available flats had been occupied (England 1976: 17). It took nearly a decade, however, before the area's vast resources could be exploited in an efficient and profitable manner. Owing to the lack of modern sawmills in the town and tramways to transport logs from the forests to the township, timber-cutting

Fig. 5. 'Transporting timber, Coff's Harbour'. Photograph from "The North Coast District" (Sydney: Government Printing Office, c1905). Reproduced courtesy of the State Library of New South Wales. Call no. X981.8/5A1. Frame no. a4342032.

remained a laborious task, with hand-cut logs hauled to the jetty by bullock team (Fig. 5). As haulages of over six miles were not viable, "the great hardwood timber reserves of the hinterland remained largely untapped" until the early 1900s (Yeates 1990: 55). In 1902, a representative of the Forestry Department visited Coffs Harbour and noted that "magnificent belts of tallow-wood, ironbark, and pine were in the vicinity" of a site allocated for a sawmill in the town centre. In the view of the representative, it was highly probable that "within a few years Coff's Harbour would be one of the most important timber centres on the north coast, owing to the shipping facilities, and

10

Proc. Linn. Soc. N.S.W., 138, 2016

Fig. 6. 'Nicholl's saw-mills, Coff's Harbour'. Photograph from "The North Coast District" (Sydney: Government Printing Office, c1905). Reproduced courtesy of the State Library of New South Wales. Call no. X981.8/5A1. Frame no. a4342034.

to the presence of untapped virgin forests" (*Evening News* 1902).

The first sawmill in Coffs Harbour opened in 1898 on the north side of the jetty, but for reasons unknown it was relatively short-lived and closed by 1902 (Yeates 1990: 64). In 1903 a major mill opened on the present site of Coffs Harbour High School (Fig. 6). (The location of this site is shown in Fig. 10 and modern views are shown in Appendix 2.) By 1906, there were three sawmills in the area – two in the Coffs Harbour town centre and one in Coramba (England 1976: 18). The growing industry increased the value of town lots, which were "readily snapped up with keen competition" (*CRE* 1905). As a journalist visiting Coffs Harbour in 1905 observed, "Selections have been taken near Coff's Harbour for timber alone" (*CRE* 1905). The activity appears to have been so rapid that, upon visiting the area the following year, a *Sydney Morning Herald* reporter was moved to remark that "the forests from which supplies are drawn [...] are now almost denuded of suitable timber" (*SMH* 1906). He appears, however, to have been referring to reserves very close to the township, for he notes that "the great forests extending on all sides", spanning "hundreds of square miles", contain "timber of every variety of hardwood, of a vastly superior kind" (*SMH* 1906). Yet these forests are not without interference:

Here is a magnificent ironbark, with the deadly mark of the ringbarking axe, and there

is the remnant of a flooded gum or blackbutt tree, from which, probably, 500 9ft rails were split, before the first fork was reached. More often the tree is standing ringbarked, bereft of life and bark, a gaunt, unlovely giant, with bare limbs extended heavenward, as though invoking a curse on its destroyer. (*SMH* 1906)

This practice was a prominent and consistent feature of the landscape: as one journalist observed in 1905, "On quite a number of selections large flooded gums, denuded of their foliage and smaller limbs, stand on the land" (*CRE* 1905). Three years later, a man from Pennant Hills visited Coffs Harbour and Coramba, and reported his observations to the *Cumberland Argus* (1908). The paper noted that, in his view, "no one makes the slightest endeavour to grow anything for domestic consumption"; instead, extractive industries took precedence in the area: "Another matter which struck him was the ruthless manner in which all settlers destroyed trees. [...] Trees which would be for ever a lasting ornament to the lands are subjected to the ring-barker's axe, without a thought." The strength of the industry was such that it even attracted foreign investors, with a South African company obtaining a lease of 60 acres including sites for a sawmill and a tramline in 1909 (*CRE* 1909b).

With the provision of decent roads and the emergence of the British Australian Timber Company tramline, which extended from the jetty to Bucca Creek by 1908, clearing became far more extensive. In particular, the tramline enabled the transportation of logs from the north-western outskirts of Coffs Harbour to the company's mill in town, thereby facilitating the efficient clearing of this area. From 1908 to 1912, the line was extended as successive areas were logged (Yeates 1990: 65). As a result of the timber industry, Coffs Harbour became "the busiest port on the far north coast of the state", with an annual average of 399 ships entering from 1909 to 1924 (Coltheart 1997:13). Simultaneously, the growth of the paspalum industry, which accelerated from 1900, introduced further changes to the landscape. The predominant approach to growing paspalum – a newly-introduced genus of the grass family –

Fig. 7. 'Hoschke's farm, Orara River'. Photograph from "The North Coast District" (Sydney: Government Printing Office, c1905). Reproduced courtesy of the State Library of New South Wales. Call no. X981.8/5A1. Frame no. a4342037.

required all hardwoods in the area to be ring-barked and all existing scrub to be felled and burnt off before the seed could be sown (Yeates 1990: 59). The result was a landscape of tall grass "growing splendidly, as it does all over this district, killing almost everything else" (*Goulburn Evening Penny Post* 1908).

The extent of the clearing by the early 1900s is apparent in photographs of three farms on the Orara River: the first identified as belonging to the Hoschke family (Fig. 7), the second to the McLeod family (Fig. 8), and the third to John Cochrane (Fig. 9). The location of these sites is shown in Fig. 10 and contemporary views of these sites are shown in Appendix 1. These photographs show that clearing was confined to the river edges on level ground, while the forest on the slopes, which was worthless from a farming perspective, was left relatively intact. In all three photographs, stumps and ringbarked trees are all that remains of the original wilderness on level ground. Drawing on recent Koala surveys (Lunney et al. 1999a), we can assume that the cleared forest would have constituted core Koala habitat.

Furthermore, three panoramic views of Bellingen (Figs 11-13) show that the transformation of the landscape for agricultural purposes was well underway across the broader district. Remnant forest is visible in these photographs, though it appears to have been heavily ringbarked in Fig. 11. Fig. 11 also shows the intensive clearing on the flat fertile lands of Bellingen and the efforts expended in forging roads.

In contrast to the farms on the outskirts, photographs of the original Coffs Harbour town centre show that certain areas retained ample vegetation and, in some areas, Koala habitat. In particular, the village known as 'Brelsford', which existed within Coffs Harbour's original boundaries, and was later renamed 'Coff's Harbour', contained significant Koala habitat. A photograph of the village (Fig. 14), taken in 1903-1905, shows that comparatively heavy vegetation still survived in suburban areas, with the distinctive timber houses nestled among the trees. Although we cannot identify specific tree species in this photograph, we can assume that these trees constituted high-quality Koala habitat, as modern

12

Proc. Linn. Soc. N.S.W., 138, 2016

Fig. 8. 'M'Leod's farm, Orara River'. Photograph from "The North Coast District" (Sydney: Government Printing Office, c1905). Reproduced courtesy of the State Library of New South Wales. Call no. X981.8/5A1. Frame no. a4342038.

Fig. 9. 'Cochrane's farm, Orara River'. Photograph from "The North Coast District" (Sydney: Government Printing Office, c1905). Reproduced courtesy of the State Library of New South Wales. Call no. X981.8/5A1. Frame no. a4342039.

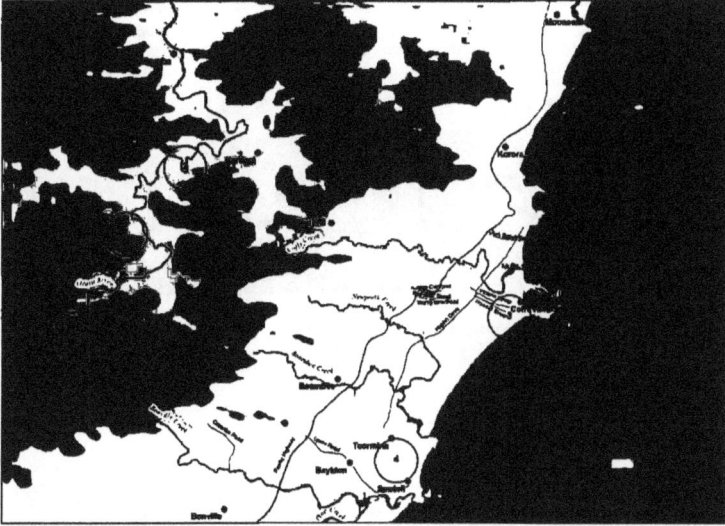

Fig. 10. Historical location map of Coffs Harbour. This map of Coffs Harbour is based on a digital elevation model, which shows that the coastal strip of Coffs Harbour is low-lying and then quickly rises away
from the coast, with valleys and rivers shown starkly by the shading. The Great Dividing Range comes
closer to the coast at Coffs Harbour than elsewhere in NSW, lending its distinct configuration to the land
of this Local Government Area. On the Orara River are sites 1 and 2 in circles: 1 is the ASD40 1, and 2 is
ASD40 2, i.e. the original sites of Hoschke's and McLeod's farms in 1, and Cochrane's farm in 2. These
circles are given here to help interpret both historical and contemporary photos of the Orara Valley by
giving the locations relative to both Coffs Harbour town and jetty (circle 3 showing the original location
of Nicholl's saw mill, ADS40 3), and the shape of the landscape. The two thick lines running from circle
3 indicate the original timber transportation routes. Circle 4 indicates the location of the 2009 aerial
photograph (Fig 26). This map also includes place names mentioned in the text.

Fig. 11. 'Part of town and North Arm of Bellinger River, from Mark's Hill'. Photograph from "The North Coast District" (Sydney: Government Printing Office, c1905). Reproduced courtesy of the State Library of New South Wales. Call no. X981.8/5A1. Frame no. a4342018.

Fig. 12. 'Rigney's Farm, near Bellingen'. Photograph from "The North Coast District" (Sydney: Government Printing Office, c1905). Reproduced courtesy of the State Library of New South Wales. Call no. X981.8/5A1. Frame no. a4342012.

Fig. 13. 'Road scene, near Bellingen (South Arm)'. Photograph from "The North Coast District" (Sydney: Government Printing Office, c1905). Reproduced courtesy of the State Library of New South Wales. Call no. X981.8/5A1. Frame no. a4342010.

Fig. 14. 'Village of Brelsford, Coff's Harbour'. Photograph from "The North Coast District" (Sydney: Government Printing Office, c1905). Reproduced courtesy of the State Library of New South Wales. Call no. X981.8/5A1. Frame no. a4342035.

analyses of the Coffs Harbour region show that the area contains extensive Koala habitat, much of it primary habitat (Lunney et al. 1999a). (Modern views of this site and surrounding areas are shown in Appendix 3.) Moreover, as clearing was restricted next to the river edges due to the high risk of flooding along Coffs Creek, we can also reasonably assume that Koalas were present at Coffs Creek. In 1900, the "land adjacent to the township, on the creek" was described as "occupied, but little cleared" (*Raleigh Sun* 1900), and by 1906 the northern side of the Creek had few settlements (Fig. 15). According to the recollections of a former resident who lived at Coffs Harbour and Bucca Creek over the period 1896-1901, "Birds and animals abounded in the bush and along the river banks" (*CHA* 1950). He names "koalas, wallabies, kangaroo rats and kangaroos" as among the animals he remembers seeing at this time. Another article supports this claim, observing that "There is a 'call of the wild' in the air of this district for [...] the wallabies and paddymelons flourish though the

sawyer haunts every woodland" (*CRE* 1909a). It is difficult, however, to determine whether the Koala was abundant: a postcard from 1907 (Fig. 16), featuring the Koala as part of a montage of images of the South Arm of the Bellinger River, indicates that the Koala was sufficiently well-known in the region to be considered representative of its fauna, but provides no further clues. Secomb (1986: 21), a long-term resident of Coffs Harbour, notes in his history of the area that among local men's responsibilities at this time was "sho[oting] the koalas to feed the dogs", but provides no source for his claim.

There is little evidence that Coffs Harbour participated heavily in the fur trade in this period, even during the depression of the early 1890s, when the trade presented a valuable source of income (Diarmid 1903). At this time, kangaroos and wallabies remained the most highly valued skins, with the Tanners and Curriers Association recommending that kangaroo farms be set up to ensure the continued supply of skins to America and Europe (*CRE* 1893).

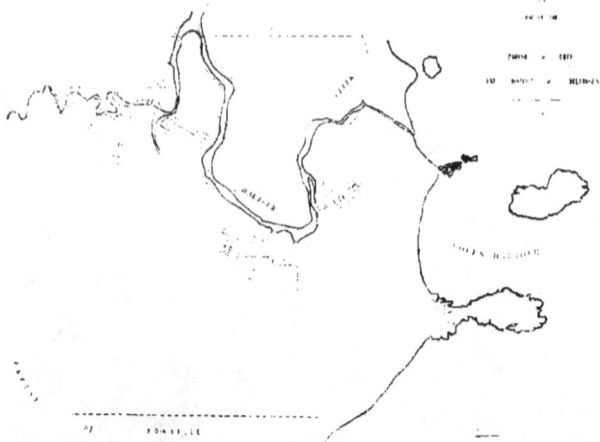

Fig. 15. 'Map of of the town of Coffs Harbour, and suburban lands, Parish of Coff, County of Fitzroy, Land District of Bellingen', 1906. Reproduced courtesy of the State Library of New South Wales. Call no. a9556001.

Fig. 16. Postcard, 'Greetings from South Arm', May 1907. The small tree growing out of a large stump in the centre of the photograph shows the considerable size of the trees that occurred near the coast, near the mouth of the Bellinger River. Reproduced courtesy of Sheila and Michael Pegum.

Proc. Linn. Soc. N.S.W., 138, 2016

17

KOALAS IN COFFS HARBOUR

Table 1: Numbers of furred skins sold through Sydney markets. The koala is shown with two other species for a point of comparison: the red kangaroo *Macropus rufus* and the brush-tailed rock-wallaby *Petrogale penicillata*. += Figures only available for first half of year; * = figures only available for second half of year. Numbers were collated by Brad Law and drawn from the Sydney Wool and Produce Journal and the Sydney Wool and Stock Journal.

SPECIES	1891*	1892	1893*	1894	1895	1896+	1897	1898	1899+
Red K	44838	141177	34856	23306	91563	35697		176862	335234
Rock W	42154		13422	9770	10317	2656		19382	9185+
Koala	57208	113629	35464	9588	31744	22563		139136	266535

The numbers of Koala skins on the Sydney market fluctuated throughout the 1890s (Table 1) and peaked after 1900. From 1891-1899 inclusive, the recorded sales of Koala skins through Sydney markets totalled 675,867 (Table 1). In contrast, 600,000 Koala skins were reportedly exported to London in 1902 alone (NSW Native Animals Protection Bill 1903). In 1904, according to fur broker Winchcombe, Carson & Co., Koala skins, alongside those of kangaroo and wallaby, were in "unlimited demand" as recorded in The Sydney Stock and Station Journal (*SSSJ* 1904). The following year, they reported that with regard to "kangaroo, wallaby and bear there are not nearly enough to go round" (*SSSJ* 1905).

In view of the demand for their fur, it is unsurprising that Koalas declined rapidly at the turn of the twentieth century in New South Wales. One commentator observed that "native bears are dying out very fast in some districts. I have seen them lying about the bush day after day" (Bellingham 1900). After Koala populations across Australia contracted what was called an ophthalmic disease in 1900-1903, leading to a reduction in numbers (Le Souef and Burrell 1926: 292; Troughton 1948: 136), the Koala was listed as a protected species in New South Wales in December 1903 (NSW *Native Animals Protection Act* 1903). Noting that in New South Wales "bears are nearly exterminated", one report commented that "it is questionable whether the Act passed recently in New South Wales [...] is not too late to accomplish its purpose" (*The Queenslander* 1905). In 1906, Koalas sold through Sydney by Dalgety & Company Limited were referred to as "Queensland Bears", a reflection of the scarcity of New South Wales koalas (SSSJ 1906). In 1910, a letter from a Macleay resident appeared in the *Sydney Morning Herald* confirming this view, making specific reference to the Coffs Harbour region:

I have made personal enquiries of surveyors, trappers, and men who spend their days in the bush. All tell the same tale – the bear is not to be seen. [...] The kangaroo, the wallaroo, the bear, and opossum have comparatively disappeared from the mountainous country that runs between the seaboard and New England on the North Coast, where a few years ago they were regarded by some people as a nuisance, so numerous were they. (*SMH* 1910)

But if Koalas declined in the Coffs Harbour area in this period, exactly what caused their demise? Examining export figures published in regional newspapers allows us to gauge the relative significance of the fur trade to Coffs Harbour at this time. While steamers called at the port regularly, with three separate lines loading cargo for Sydney by 1894 (*CRE* 1894), skins were not a major export. Though the town-specific reports in the *The Clarence and Richmond Examiner* were irregular, the available reports for Coffs Harbour indicate that the skin trade was inactive in the area in the mid-to-late 1890s. An annual export report for 1895 lists timber as Coffs Harbour's primary export, and the list of minor exports does not include skins (*CRE* 1896). Annual reports for 1897 and 1898 also reaffirm the importance of the timber industry in the area, seconded by maize, and skins are absent from their precise lists of exports (*CRE* 1898; 1899). Interestingly, however, skins are listed among the exports for 1901, with the annual total exported from Coffs Harbour "25 bales" (*CRE* 1902).

Monthly breakdowns of the exports from each area of the North Coast allow us to place the Coffs Harbour skin trade in a regional context. In these reports, Coffs Harbour and Woolgoolga (a town to the north of Coffs Harbour and within the current Local Government Area) comprise a single district which, when compared to other districts, consistently proves a minor contributor to the skin trade. Interestingly, the Bellinger district also proves relatively minor. For the month of December 1906, skin exports from Coffs totalled "2 bags/bundles", compared to 38 for the Clarence River, 21 for the Richmond River and 1 for the Bellinger districts (*CRE* 1907). For July 1908, Coffs exported "5 bags/bundles", while the Clarence exported 90, the Richmond River exported 61 and the Bellinger exported 5 (*CRE* 1908a). Though the figures for the surrounding districts fluctuate, Coffs remains generally consistent: for August 1908, it exported "3 bags/bundles", with the Clarence, Richmond River and Bellinger districts exporting 79, 86, and 14 respectively (*CRE* 1908b). It is important to note that, as these figures were presented in the aggregate, we cannot identify the specific type of skin being exported. Regardless, Coffs Harbour's low numbers indicate that the trade in the area was minimal. In view of this, advertisements placed in *The Coffs Harbour Advocate* by fur/skin brokers in 1910 (*CHA* 1910a,b) signal an attempt to meet increasing demands for fur at a time when the Koala, alongside other marsupials, was widely perceived to be declining across New South Wales, rather than evidence of an active trade in Coffs Harbour.

Though conservationists at this time generally placed the blame for the Koala's decline squarely on the fur trade, campaigning for hunting restrictions ranging from closed seasons to absolute protection (Moyal 2008), the effects of the trade on Koala populations appear to have been variable across districts. With regard to Coffs Harbour, it is especially telling that, in the mid-to-late 1890s, timber exports consistently rose while the skin trade remained inactive. In this period, timber exports increased from 106,500 feet in 1895 to 480,510 feet in 1898 (*CRE* 1896; 1898; 1899) [here, 'feet' is assumed to denote super feet, with a super foot being a unit of volume of timber in the imperial system of 1 foot x 1 foot x 1 inch]. There is little evidence to suggest that Coffs Harbour participated in the fur trade prior to 1901, and if we take the generality of *The Clarence and Richmond Examiner*'s commercial reports into account, no evidence. Rather, the growth of Coffs Harbour's timber industry, the speed of land clearing, and the widespread practice of ringbarking would have exerted a far greater effect on the local Koala population than the trade in marsupial furs and skins. Furthermore, it is probable that the departure of many of Coffs Harbour's men for the First World War diminished this already minor trade.

While we can safely conclude that the timber industry, clearing, and ringbarking would have led

to considerable habitat loss and fragmentation, it is more difficult to ascertain to what extent the Koala population declined over this time. In view of the absence of a population baseline, we can draw few definitive conclusions. However, in light of the available evidence, it appears unlikely that the Coffs Harbour Koala population was rapidly and severely reduced over the period 1890s-c1920s from an initial considerable size at European settlement, such as occurred in the Bega District or Port Stephens Koala populations (Lunney and Leary 1988; Knott et al. 1998). Most likely, the Coffs Harbour Koala population was reduced in numbers as habitat was lost and fragmented. At the close of this period, Koalas remained present in the Coffs Harbour area, particularly the town centre and near waterways, but were not especially plentiful.

1920-1950s

We have been able to identify very little material specifically regarding Koalas for this period of Coffs Harbour's history. Coverage of animals in *The Coffs Harbour Advocate* over this period consistently focused on those animals considered pests, such as flying-foxes (*CHA* 1920), opossums (*CHA* 1930), and wallabies (*CHA* 1940). However, beyond media analysis, other sources provide us with evidence of both a present Koala population in the Coffs Harbour area, particularly the town centre, and of potential threats to this population.

Consistent with the probable presence of Koalas in the Coffs Harbour town centre in the early 1900s, available evidence strongly suggests that Koalas remained present in the town centre and surrounding area from the 1920s to the late 1950s. A long-term resident of Coffs Harbour, born in 1905 and interviewed for the Coffs Harbour 'Voices of Time' project in 1987, recalls seeing Koalas in the town in the late 1920s and early 1930s (Mayers 1987). Evidently, they were still present in the late 1930s: an article, entitled "Koala on Road" and written by a visitor to Coffs Harbour, describes seeing "a large grey koala bear" crossing "one of the main roads leading from Coff's Harbour, only about three-quarters of a mile from the town" (*The North Western Courier* 1939). The land bordering Coffs Creek appears to have retained considerable Koala habitat in the 1920s and 1930s, with photographs taken in this period showing scrub and bushland surrounding the creek (Figs 17-18). Furthermore, a photograph published in Yeates (1993: 8) of a longstanding Aboriginal campsite on land bordering the south bank of Coffs

Proc. Linn. Soc. N.S.W., 138, 2016

19

Creek testifies to the persistence of core Koala habitat in the area in the late 1930s. A photograph of another Aboriginal campsite near the creek, situated in bush closer to the cemetery, shows that the area contained Koala habitat in the late 1950s (see Yeates 1993: 204). Further photographs show that the township remained surrounded by scrub and bushland from the 1930s through to the late 1950s (Figs 19-20; Yeates 1993: 263). As tree planting programs – part of a town 'beautification' initiative – only began in the mid-1950s, we can safely assume that the vegetation featured in these photographs is far older, though the exact age cannot be determined.

The broader area surrounding the township appears to have retained a number of older trees despite comprehensive clearing. Yeates' comprehensive local history alerts us to the presence of "a Flooded Gum with a girth of 23 feet, and standing 215 feet high beside the road" in the Bruxner Park Flora Reserve in 1961 (Yeates 1993: 324). It appears, however, that such a large tree was a rarity in the town centre at this time, for he notes that it "was admired and often photographed by those who saw it". In the forests surrounding Coffs Harbour, a number of older blackbutt trees survived the extensive clearing and logging of earlier decades, including one felled in 1950 in the Upper Orara State Forest and measuring 100 feet x 16 feet centre girth

(Yeates 1993: 62). Another, processed by sawmillers Seccombe and Forsythe in 1960, was "delivered in three sections, the middle one of which was 18 feet long, 21 feet 6 inch girth at the middle, and assessed at better than 16 tons in weight". Seccombe, a veteran of the Coffs Harbour timber industry, described it as "just about the biggest he had seen in a lifetime with timber" (Yeates 1993: 273). It had been cut from the Never Never State Forest, now part of Dorrigo National Park.

It is highly unlikely that the fur trade constituted a potential threat to the Koala population of Coffs Harbour in this period. Since the early 1920s, the trade in Koala skins had been centred on Queensland, as Koala populations in the south-eastern states had declined significantly (Marshall 1966; Moyal 2008). Moreover, as we have seen, Coffs Harbour did not have a strong history of participation in the trade. There is no evidence to suggest that Coffs Harbour locals turned to hunting Koalas – or other marsupials – during the Great Depression of the late 1920s and early 1930s, with residents turning to casual jobs such as packing bananas, selling fish caught in local creeks to shops, and tomato-picking at this time (Yeates 1990: 204). Furthermore, human population trends, based on data compiled by Yeates (1993: 338), show a marked decline during the 1920s and Great

Fig. 17. 'View down Coffs Creek towards town, Coffs Harbour, N.S.W', by Peter Jensen, 1924. Reproduced courtesy of the Coffs Harbour Library and Coffs Harbour Regional Museum. Accession no. 07-8297.

Fig. 18. 'View of Coffs Creek, Coffs Harbour, N.S.W.', c1925. Reproduced courtesy of the Coffs Harbour Library and Coffs Harbour Regional Museum. Accession no. 07-9028.

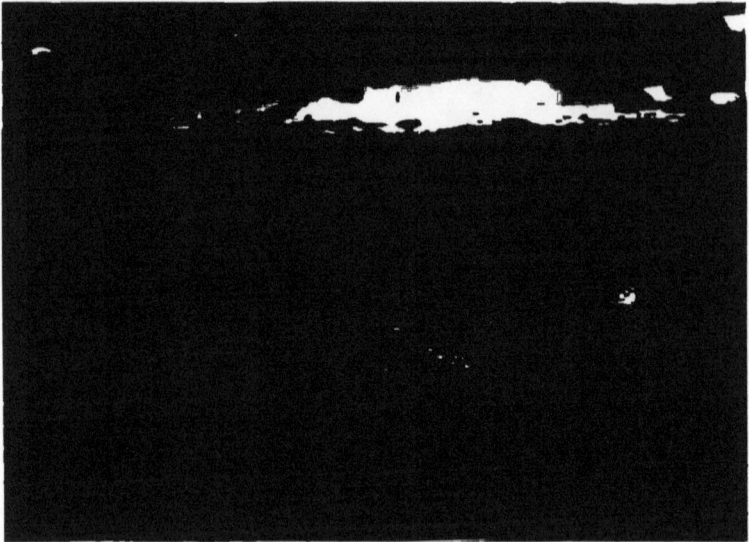

Fig. 19. 'View from the Jetty area, looking back towards Coffs Harbour township and Red Hill', c1920s. Reproduced courtesy of the Coffs Harbour Library and Coffs Harbour Regional Museum. Accession no. 7-1927.

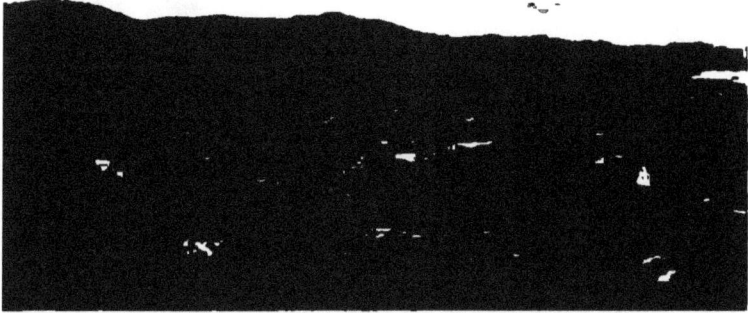

Fig. 20. 'View of the Jetty area, including the Butter Factory, Memorial Theatre and High School, Coffs Harbour', c1940s. Reproduced courtesy of the Coffs Harbour Library and Coffs Harbour Regional Museum. Accession no. 07-1909.

Depression, with the population only returning to pre-1915 level in the 1940s. Over the late 1940s to 1964, the population steadily increased. Given the slow pace of this demographic shift, we can assume that population growth did not constitute a major threat to the Coffs Harbour Koala population in this period.

Similarly, it is questionable whether bushfires presented a threat to the Koala population of the area. Local and regional newspapers show that there were frequent fires in the area over the late 1930s to mid-1950s. In November 1936, bushfires spread through the hardwood forests of the North Coast, including Tanban, Ingalba, and Barraganyatti State Forests, all south of Coffs Harbour. The damage in the Coffs Harbour area itself was described as "extensive", affecting a number of banana plantations and the "scrub country" on the Dorrigo (*SMH* 1936a). The following month, another wave of bushfires ravaged the townships surrounding Coffs Harbour, including Boambee and the Upper Orara, with firefighters forced to "drive through several miles of blazing scrub" to reach the township of Orara (*SMH* 1936b,c). Bushfires struck again in the late 1940s near the Coffs Harbour aerodrome. In 1951-53, successive bushfires broke out in the area. Fires spread through "thousands of acres of scrub around Coffs Harbour" in late 1951, with an aerial view of the area captioned "smoke from the fires rose over 6,000 feet" (*SMH* 1951c). One blaze "destroyed" 300 acres of timber on Boambee Mountain, south of Coffs Harbour (*SMH* 1951a). One report noted that in Coffs Harbour "fresh fire outbreaks are occurring hourly", quoting the District Forester as stating that "at one period this morning 22 separate fires were burning" in his district, with the worst outbreaks "concentrated in the Conglomerate, Wedding Bell[s] and Orara State Forests" (*SMH* 1951b). In 1953 another wave of fires swept the district, particularly affecting Boambee and Bonville (*SMH* 1953).

However, fire history maps, prepared by the Office of Environment and Heritage, allow us to conclude that fire has been a relatively minor matter over the last 75 years for the Koalas occupying the forested land in and surrounding Coffs Harbour. As shown in Fig. 21, fires have been concentrated on the northern and north-western border and the south-east corner of the Coffs Harbour Local Government Area (LGA). Fig. 22 shows that the majority of the fires have been wildfires, as opposed to prescribed burns, and that these have occurred on the fringes of the LGA. Prior to 2005, the northern tip, consistently the site of wildfires, was not included in the Coffs Harbour LGA. Most importantly, we must take into account recent research which shows that Koalas can re-occupy burnt bushland within months of a fire, and breed in it within a year (Matthews et al. 2007).

22

Proc. Linn. Soc. N.S.W., 138, 2016

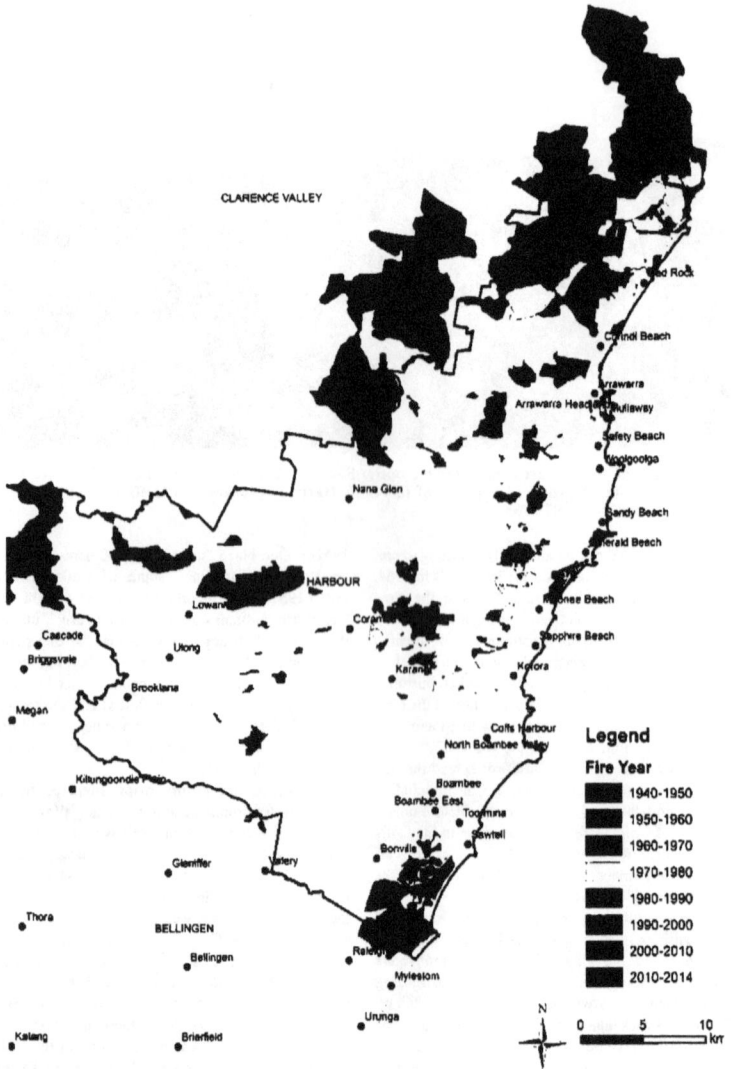

Fig. 21. Fire history of the Coffs Harbour LGA (current boundary), 1940-2014. Produced by the NSW Office of Environment and Heritage and stored in corporate data layers.

Proc. Linn. Soc. N.S.W., 138, 2016

23

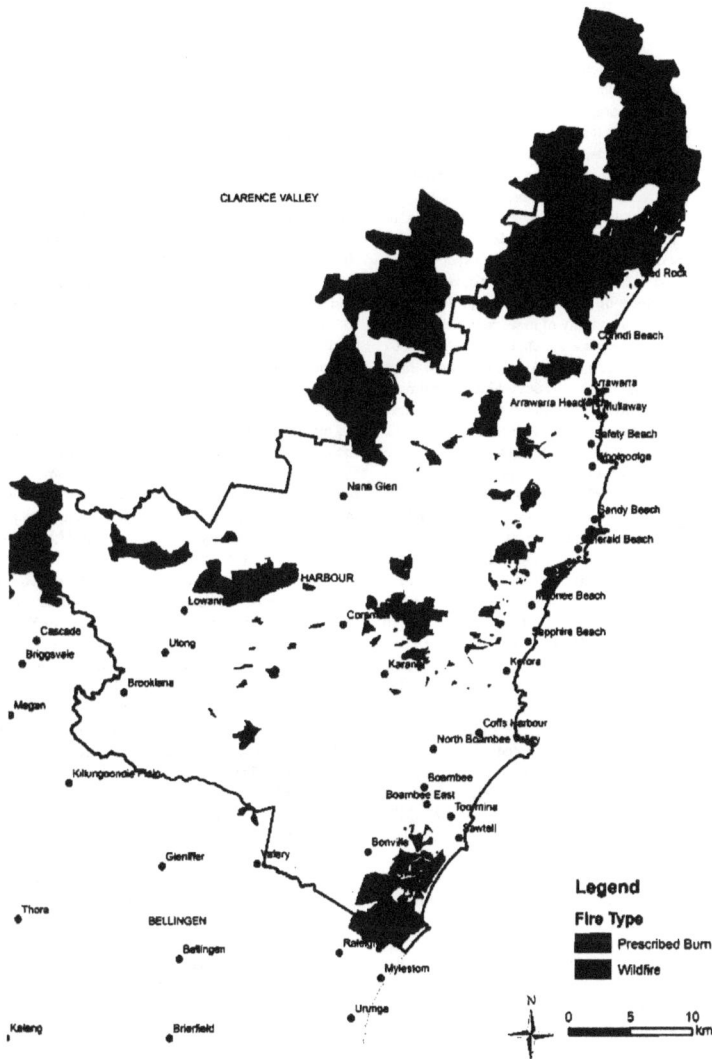

Fig. 22. Fire history of the Coffs Harbour LGA (current boundary), 1940-2014, showing prescribed burns and wildfires. Produced by the NSW Office of Environment and Heritage and stored in corporate data layers.

The fires would have temporarily affected the status of the Koala populations in the forests surrounding Coffs Harbour, insofar as the Koalas inhabiting the areas where the fires occurred would have been killed. However, the long-term impact of the fires on the presence of these populations can be considered negligible due to the rapid rate of recovery of the forest as Koala habitat. As the surrounding unburnt forests were extensive in area, they would have provided a crucial source population for the rapid recolonisation of the burnt areas as these areas recovered.

In contrast, the timber industry continued to present a threat to the Koala population of the area over this period. Although the industry experienced a decline in the late 1920s and early 1930s, largely due to the importation of timber from the United States and the growing popularity of new industries such as banana farming, it regained strength in the mid-1930s (Yeates 1990: 191, 226, 243). As a result of the sharp rise in banana production, the demand for case timber grew, and by 1938 Coffs Harbour had 5 case timber mills and 5 general sawmills (Yeates 1990: 228). Though hardwood remained the key export from the area, Flooded Gum was planted in an effort to maintain the supply of case timber (Yeates 1993: 61-62). The Second World War brought with it a fresh demand for hardwood, with large quantities of blackbutt sent to New Guinea for use by the American Army (Yeates 1990: 243). While exports slumped after 1945, due to the unavailability of coastal shipping vessels, an acute housing shortage in Coffs Harbour generated high local demand (Yeates 1993: 5). In 1949, shipments of timber totalled 11.6 million super feet over nine months, with the chainsaw replacing older cutting methods (Yeates 1993: 62).

During the 1950s, locals developed a number of measures which, though not intended to conserve local fauna, may have inadvertently assisted the continued presence of the Koala in the area. In 1952, presumably as part of its town 'beautification' program, the Coffs Harbour Urban Committee banned the removal of existing trees in the township, and granted exceptions only for "dangerous specimens" (Yeates 1993: 110). The same year, a Forestry Commission representative who had worked in the area since 1912 set the export trade on a path of reform, declaring that by meeting market demands for only the best poles and piles of specified species, the Coffs Harbour area would be denuded of the best timber (Yeates 1993: 151). Furthermore, he publicly stated that "the forests were deteriorating at a faster rate than Nature was able to replace them" (Yeates 1993: 151). Despite this prescient observation, record quantities of timber were exported from Coffs Harbour over the period 1956-

1959 – 80 million super feet in 1958 alone (Coltheart 1997: 15; Yeates 1993: 272). In 1958, in order to meet the demands of the export industry, a local group acquired 1,750 acres of degraded farmland for the purpose of establishing eucalypt and pine plantations. After successive purchases of surrounding tracts, the plantation companies eventually amassed 40,000 acres of land on both sides of the Pacific Highway (Yeates 1993: 273).

A representative of the NSW Office of Environment and Heritage in Coffs Harbour and specialist in Koalas, John Turbill (pers. comm. 2014), states that many of the plantations surrounding Coffs Harbour and Bellingen would have contained remnant forest along creeks and road edges. These remnants would have enhanced the quality of the Koala habitat within the plantation, increasing the likelihood of these plantations serving as habitat corridors for local Koalas. Additionally, as these plantations grew, they would become progressively more likely to attract Koalas. Some Koalas would, over time, include the plantation within their home ranges, which can encompass both plantation and non-plantation forest. Conceivably, a Koala could come to spend some or all of its time in the plantation and adjacent old growth forest. Indeed, Smith (2004) showed that Koalas occurred at low density in the plantations within Pine Creek State Forest (18 km south of Coffs Harbour) at the time of his research in the 1990s. Smith (2004:591) reports that Koala density varied from one Koala per 50 ha in plantation forest to one Koala per 9 ha in high quality native forest. The importance of managing Koalas within Pine Creek State Forest is evident in a Koala Management Plan (State Forests 2000), and Newman and Partners (1996) give a more extensive history of forestry management in this forest. In 2003, the bulk of the prime Koala habitat of Pine Creek State Forest was transferred to Bongil Bongil National Park, which had been established in 1995 [Bongil Bongil National Park is 4233 ha and Pine Creek State Forest is 3511 ha as of February 2015]. With regard to the plantations bordering the Pacific Highway, Koalas have been regularly sighted crossing the Highway. Lassau et al. (2008), in a study aiming to ameliorate the effect of roadkill on Koala populations at Bonville (within the Coffs Harbour LGA), show that fencing had proved an effective barrier to Koalas crossing the Highway.

Relative to the extensive areas of native forest within the Coffs Harbour LGA, however, plantations are a minor feature of the forest estate in the area. Fig. 23 shows that plantations, although extensive, lie to the west and south of the Coffs Harbour LGA. The plantations at the very northern tip of the LGA are

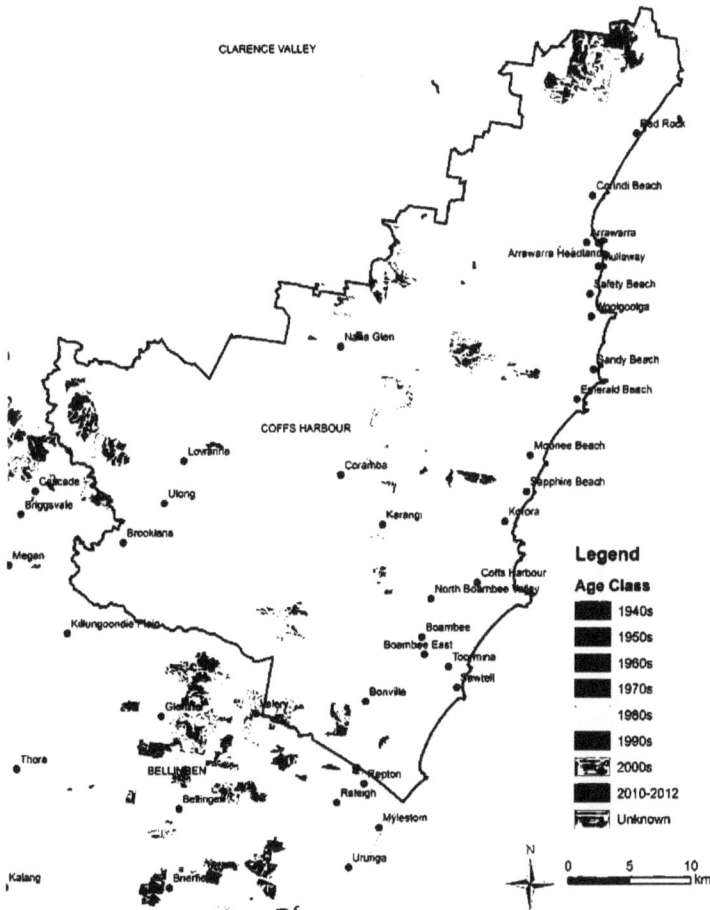

Fig. 23. Plantation history of the Coffs Harbour LGA (current boundary), 1940-2012. Produced by the NSW Office of Environment and Heritage and stored in corporate data layers.

recent, and this northern tip was not in the LGA until 2005. Given that the Koala population of the Coffs Harbour LGA is concentrated in the south-eastern sector (Lunney et al. 2000), and the larger plantations are located to the south and west of the LGA border, particularly in Bellingen LGA, it is apparent that plantations are not a major factor in the current distribution of the local Koala population. However, we can surmise that the creation of plantations on reclaimed farmland within the Coffs Harbour LGA would have increased the area of low-density Koala habitat.

THE DEVELOPMENT BOOM AND THE
EMERGENCE OF A CONSERVATION ETHIC:
1960-2000

26

Proc. Linn. Soc. N.S.W., 138, 2016

On a statewide basis, Reed et al. (1990) found that the Koala population of north-coast New South Wales remained constant in the postwar decades, whereas losses occurred on the southern half and the western fringe of its former distribution. Focusing on Coffs Harbour allows us to form a refined picture of change that is not discernible from a statewide overview. For our purposes, it is particularly important that, whereas other coastal areas had undergone development earlier in the century due to settlement patterns, Coffs Harbour experienced significant human population growth only since the early 1970s.

Following the revitalisation of local business in the late 1950s, large-scale development began in the early 1960s with the launch of several major subdivisions (Yeates 1993: 247). Amidst increasing coverage of conservation issues, particularly those concerning fauna, in local media (*CHA* 1960a,b,c,e), a representative of the NSW Fauna Protection Panel stated that the Panel was "extremely concerned by the reduction in the numbers of eucalypt trees" in Coffs Harbour, fearing that this would threaten the "numerous koala colonies" it had identified in the area (*CHA* 1960d,b). Despite such warnings, development proceeded. The Jetty area, Coffs Harbour town centre, and the 'Brelsford' district underwent extensive development over the late 1960s to 1980s, as town planners sought to accommodate an increasing population, establish industrial estates and associated road infrastructure, and establish the area as a tourist destination. Over these years, development projects consistently received positive coverage in *The Coffs Harbour Advocate* (*CHA* 1970a,c; 1980a,b), with front-page criticism aimed at preserving the tourist hub of the area and not its fauna (*CHA* 1980c). An atypical letter to the paper in mid-1970 prefigures the environmentalist opposition to unchecked development that would dominate the public debates of the late 1980s:

> Is there any thought given to the plight of the koalas in all this "progress" which is taking place around Coffs Harbour. There are more koalas in this area than people realise. Does the land developer or bulldozer driver check the gumtrees before commencing to destroy the koalas' environment, or is it left to "chance" that the koalas will get out of the way in time before it is snatched from underneath him, plus into the bargain, face death or be maimed. [...] This clearing is going on every day of the week. One driver told us that if they would only go to the side, instead, they keep moving in front of

the machines all the time. [...] A cat gets more protection than a koala when he trespasses on private property. The hazards of bushfires, the trigger-happy rifleman and wild dogs are more than enough for them to put up with. (*CHA* 1970b)

Although conservation issues continued to receive occasional attention (*CHA* 1970d,e), this was clearly the minority view. As the development of Coffs Harbour expanded south, beyond the town centre, to envelop the coastal strip east of the Pacific Highway, the high quality Koala habitat in the area was progressively eradicated. The habitat that remained became increasingly fragmented and, as a result, exposed to threats such as motor vehicles and domestic dogs (Lunney et al. 1999a). The current management of the Coffs Harbour Koala population is an attempt to deal with the threats that arose from decades of development and, in particular, the effects of the relentless loss and fragmentation of habitat.

Comparing aerial photographs of the south-eastern sector of the Coffs Harbour LGA, south of the Coffs Harbour township, allows us to discern the speed of change in specific areas over recent decades. Three geo-referenced photographs, showing the same area over 45 years, display the shift from a largely rural landscape to one that is predominantly urbanised. In 1964 (Fig. 24), the area was characterised by large patches of native vegetation interspersed with farmland. The older settlement of Sawtell is distinguished from the surrounding lands by its cluster of houses. By 1984 (Fig. 25), it is evident that housing development is occurring in clusters, consistent with a rapidly urbanising landscape incorporating associated infrastructure such as roads, in addition to electricity and water supplies. Collectively, this infrastructure exacerbates the loss of Koala habitat and increases the threat levels of dog predation and roadkill. Large tracts in the western sector of the LGA largely remain farmland, thereby continuing to support any pre-existing Koala populations. The Koalas that are visible to the residents of the new housing estates could potentially originate from this persisting rural landscape, but their presence in the new urban areas is likely to be short-lived because these areas could not sustain Koala populations.

The development underway by 1984 has been visibly consolidated by 2009, as shown in Fig. 26. This aerial photograph displays a heavily populated landscape, marked by a density of schools, shopping centres, and housing estates, and serves as evidence of the intensity of development since 1964, particularly in the area north of Lyons Road. Patches

27

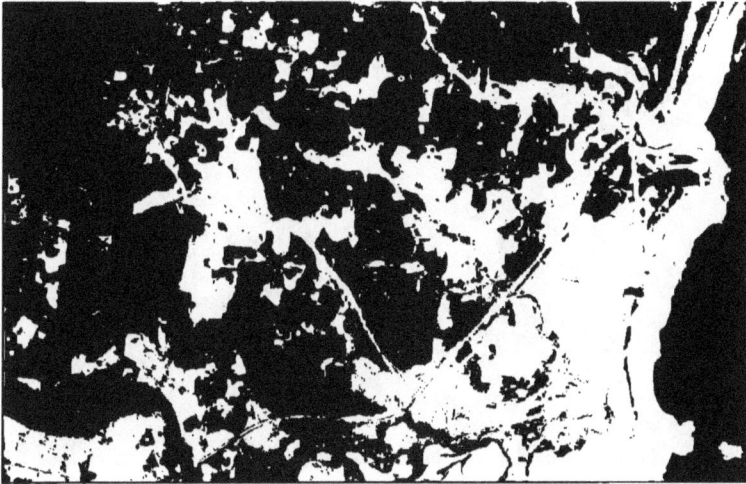

Fig. 24. Geo-referenced 1964 aerial photograph of a portion of the south-eastern sector of the Coffs Harbour LGA. This sector is bounded by the Pacific Ocean to the east, with the well-established village of Sawtell identifiable by its cluster of buildings on the coast. Bonville Creek forms the southern boundary of this photograph, and is identified by the circled area marked '4' in "Fig 10, which gives the general location of the area shown in this photograph. The road pattern is also shown in "Fig. 10".

Fig. 25. Geo-referenced 1984 aerial photograph of a portion of the south-eastern sector of the Coffs Harbour LGA. The outline of this aerial photograph exactly corresponds to that of Figs 24 and 26 to enable direct comparison.

Fig. 26. Geo-referenced 2009 aerial photograph of a portion of the south-eastern sector of the Coffs Harbour LGA. The outline of this aerial photograph exactly corresponds to those of Figs 24 and 25 (1964, 1984) to enable direct comparison. This high-resolution ADS40 photograph is much sharper than the previous monochrome aerial photographs, necessitating an element of careful interpretation to the earlier photographs of the same site. However, what is striking is the change over 45 years from an essentially rural and forested landscape to one of high-density housing with isolated patches of forest dissected by roads. Nevertheless, some Koala habitat is still visible, as are connecting links in the landscape, such as the vegetation bordering Bonville Creek, on the southern boundary of the photograph. This explains why some Koalas would occasionally be seen in urban areas.

of vegetation remain throughout the housing estates, explaining why Koalas are still occasionally seen even within urbanised areas. The density of housing and associated infrastructure indicates that this locality is more likely to see more Koala deaths than births. Comparison of these photographs reveals the speed of development in Coffs Harbour in the last three decades of the twentieth century – but the most recent stage in the long-term conversion of Koala habitat to a landscape with more threats than opportunities for Koala populations to be sustained.

The first Comprehensive Koala Plan of Management [CKPOM] in NSW (Lunney et al. 1999a), prepared for Coffs Harbour City Council and adopted in 1999, details specific examples of habitat fragmentation through development. The authors identified a distinct pattern of Koala distribution in the Coffs Harbour area, with the predominant number of records in the south-eastern sector from Moonee to Bonville (Lunney et al. 1999a Part B: 27). This area was also "the most urbanised area", characterised by "increasing urban expansion and an increasing number of road links between the business

district of Coffs Harbour and the nearby satellites of Bayldon and Toormina" (Lunney et al. 1999a Part B: 27). Furthermore, the report concluded that Preferred Koala Habitat was "highly fragmented due to coastal development and agriculture" (Lunney et al. 1999a Part B: 45). In particular, the developed area of Sawtell, Bayldon and Toormina, to the south of the Coffs Harbour town centre, was found to bisect the area of preferred habitat. Similarly, the authors identified that the Pacific Highway "generally splits the Preferred Habitat - type A on the coast from the Preferred Habitat - type B to the west" (Lunney et al. 1999a Part B: 45). The fragmentation, loss and destruction of habitat are shown in habitat map B7 (Lunney et al. 1999a Part B: 50). In addition to "clearing for urban development, bananas and grazing", the authors identified further factors which contributed to the degradation of Koala habitat in Coffs Harbour: "clearing or thinning of timber during property development, selective logging, regular burning, pollution and the proliferation of weeds" (Lunney et al. 1999a Part B: 45).

Due to the bisection of Preferred Habitat by

the Pacific Highway, Koala roadkill constitutes a persistent threat to the conservation of the Coffs Harbour Koala population. Over the period 1990-1995, Coffs Harbour WIRES was notified of 85 Koalas involved in road accidents, of which 73 (86%) died (Moon 1995). The CKPoM identified Boambee and Toormina as the worst areas for Koala road accidents, followed by Bonville, Korora and Red Hill (Lunney et al. 1999a Part B: 56). As the authors note, the available figures likely underestimate the true impact on the Koala population, due to the probability of further Koalas being hit and dying on the side of the road or later in the bush from injuries, where they are not visible to motorists (Lunney et al. 1999a Part B: 56). Human population growth is a key factor behind this problem, which has persisted. When the authors of the CKPoM conducted their 1990-91 Koala Survey, the population of Coffs Harbour was 51,520 (ABS 1991). Ten years later it had grown to 61,186 – faster than the growth rate of New South Wales (ABS 2001). Arguably, these statistics allow us to gauge the speed with which Koala habitat was progressively degraded to accommodate housing and urban infrastructure. Additionally, they contextualise the rising threats to Koala conservation associated with the human population, such as the presence of an increasing number of motor vehicles and dogs in the area (Fig. 27).

These threats, among others, were identified by respondents to the 1990 Koala Community Survey conducted in Coffs Harbour by the authors of the CKPoM. Complemented by a field survey, this survey was undertaken in order to identify Koala habitat in the Coffs Harbour LGA and to provide a firm basis for management and planning in the lands over which Council had authority [Coffs Harbour City Council has jurisdiction over private lands. This excludes Crown lands, i.e. State Forests and National Parks. At the time of the Koala Survey in 1990, Crown lands comprised 42% of the Shire]. Its methods are detailed elsewhere (see Lunney et al. 1999a, 2000). Respondents to the survey lived in all areas of the LGA, but there was a higher percentage return from the areas of Coffs Harbour, Sawtell/Bayldon/Toormina and Corindi/Woolgoolga, which are the major centres of the LGA. Koalas were observed frequently in many areas excepting Corindi/Woolgoolga, Lower/Central Bucca and Glenreagh/Nana Glen. A majority of respondents had seen Koalas in the past 12 months in the areas of Dairyville/Fridays Creek, Ulong/Lowanna, Karangi/Coramba/Red Hill and Boambee. The majority of respondents in the areas of Corindi/Woolgoolga, Glenreagh/Nana Glen and Coffs Harbour had not seen Koalas in the past 12 months. In answer

to the question "In the time you have lived in your local area has the number of koalas (a) Increased, (b) Stayed the same, (c) Decreased, or (d) Don't know?", the majority of the 1,856 respondents (75%) selected option D. Of those that did have an opinion, most (15%) said that the population had decreased, whereas only 2% of respondents stated that it had increased. This community wisdom has been shown to be effective in describing patterns of population change in Koalas (Predavec et al, in press).

Drawing on the combined results of the community and field surveys, the authors identified that the predominant number of records were in the south-eastern sector of the LGA, from Moonee to Bonville (Lunney et al. 1999a Part B: 27). Specifically, examination of the seven detailed local area maps shows that Coffs Harbour's main Koala population extended from the southern half of the Korora area, south through Coffs Harbour town area to Bayldon/Toormina and through to Boambee and the northern

Fig. 27. Dogs are a recognized threat to Koalas. This photograph was taken in 1988 in a backyard in Playford Avenue Toormina. The paddock in the background is now a housing estate. Photograph by John Willoughby.

part of the Bonville local area (Lunney et al. 1999a Part B: 27). It is possible that the Koalas present in this area reflect emigration from Koala habitat elsewhere, and that such new suburban growth areas are 'sinks' for Koalas, i.e. that the local death rate exceeds the birth rate. Additionally, the increasing human population growth in these areas introduces a potential bias in the data. It is also reasonable to speculate that as the housing estates expanded over the late 1970s through to 1990, Koalas would have been more visible, as they spent more time walking between patches of habitat and crossing roads, potentially giving a false impression of a more stable population than is actually the case. However, this distribution bias arising from visibility was mitigated by the field survey, which was independent of human population distribution.

The 1990 community survey provided a section for respondents' comments. Of the 2,018 returned forms, 1,021 (51%) contained a comment. These comments were published as a supplement to the CKPoM (Lunney et al. 1999b) and comprise an important source of perceptions data. Many respondents' comments contained observations and opinions regarding what they saw as the key threats to Koalas in the area. Development was considered to be the principal threat to the Koala population, with 90 respondents of a wide age range mentioning development in terms that convey their awareness of ecological ideas such as habitat, food source, and wildlife corridors.

"Destruction of habitat – over development of Coffs Harbour – main cause of their demise". (Male, 71, Coffs Harbour)

"There seems to be a loss of food for koalas from development such as Pacific Bay Resort". (Female, 27, Mullaway)

"Am very concerned about recent logging in the area that I saw the koala. As it was young and healthy looking I feel there must be a colony out there". (Female, 36, Coramba)

"It was a great joy to sit in the lounge and look out the window and see a koala in a tree with a baby. [...] now we see few. One only, 2 weeks ago - since removing trees which was the corridor to Bruxner Park when the Eden Park Estate was cleared". (Female, 65, Coffs Harbour)

"The majority of the respondents concerned by development mentioned specific examples

of clearing which, in their view, had exerted detrimental effects on Koala populations".

"Koalas disappeared when the land for Fitzroy Gardens and Sunbird Estate was developed" (Female, 65, Toormina)

"Koalas were plentiful near us until the Don Patterson Drive was put through their habitat. None seen since road put there". (Female, 61, Coffs Harbour)

"I have been told recently that trees are being bulldozed in the middle Boambee area for a proposed development, and that residents of that area say that koalas are coming crying to their houses in the night as their trees have been knocked down". (Female, 74, Corindi Beach)

"We live (near to) proposed Bonville Golf Course which was APM land. Since clearing commenced 2 months ago we have not seen any koalas at all, and we are concerned as to where they have gone, as there is not much bush left". (Female, 42, Bonville)

Other specific examples mentioned by respondents include Quinwell Estate (Sawtell), Pacific Bay Resort, the clearing of trees bordering the Coffs Creek tributary, and habitat destruction in Daniels Road, Coramba. Many respondents were sensitive to the connections between development and potential Koala roadkill. 21 respondents mentioned sighting a dead Koala on a road, while others displayed an awareness of potential threat:

"When we moved to Bonville koalas were frequently sighted. Now that residential areas have replaced bushland, we only frequently see koalas dead on the road". (Female, 17, Bonville)

"1981 sighting of koala 9.45pm crossing road slowly from paperbark stand east of Hogbin Drive to west side. [...] I assume that Hogbin Drive, newly made, had cut across the bear's territory". (Male, 67, Sawtell)

"Five koalas have crossed McKays Road during last 12 month period, due mainly to urban development west of McKays Road. This is a high risk area for koalas". (Male, 70, Coffs Harbour)

Respondents also mentioned a variety of other factors that in their view constituted threats to the Koala populations of their area. These included the presence of roaming dogs (48 comments), cats (16 comments), wildfires (2 comments), and flow-on effects of development such as noise and smoke (2 comments).

Most interestingly for our purposes, many respondents claimed that Koalas had successively declined over time and utilised a historical frame of reference to substantiate these claims. Of these respondents, many felt that development had been the key factor in their apparent demise. As one male respondent commented, "I have lived in the Karangi area all my life and have seen a decline in koalas mainly due to more traffic, land clearing for bush retreats and power lines" (Male, 33, Coramba).

Others observed a decline but did not attribute a reason, with one woman stating, "As a child I saw lots of koalas in this area. Our children haven't seen any" (Female, 43, Nana Glen), and another respondent commenting, "In my last 17 years I've gone from seeing a koala on an average of once a month, to now only seeing them once every two years" (Female, 27, Coffs Harbour). However, it is particularly interesting that multiple respondents observed a decline in what they perceived to be an already small population:

"Koalas were usually seen on the farm round October, where there was still plenty of natural bush. This changed when it was cleared for development in '70. Wouldn't say they were ever plentiful". (Female, 67, Sawtell)

"I am now in my 60th year. In 1937 I saw my first koala in the Conglomerate State Forest where I spent quite a few years riding horseback looking for grazing cattle. I saw another koala crossing the road one night when driving by car from Coffs Harbour at the lower Bucca turn-off. With many hours spent in the bush as a young person these are the only two I have seen in the wild". (Anon., N/A, Coffs Harbour)

A distinct consensus emerged with regard to the pattern of Koala distribution in Sawtell. While one respondent mentioned that "quite a few koalas" were present in the early 1980s between Toormina High School and Sawtell/Toormina Roads, the majority of respondents observed that despite years of residence in the area they had never, or only rarely, spotted a Koala. As one respondent commented, "Yes, we used to have koalas in our area, but they are very hard to find and always have been. Even in trees we know they were feeding in we very rarely ever seen them" (Female, 57, Sawtell).

The comments of residents who had lived in the area for three decades or more are particularly illuminating:

"My husband owned approximately 300 acres of bushland from the Lyons Railway bridge to Pacific Highway from 1946, felling timber then bulldozer, and in all those years didn't sight one koala, but saw one in Karangi Bush". (Female, 79, Sawtell)

"I always look for koalas when passing through forest areas but have only ever seen the one. I live on the edge of the Sawtell Beach scrub. I have only seen one koala in Sawtell area in my 30 years residence here. It was a fully grown one on Sawtell Reserve about 20 years ago". (Male, 89, Sawtell)

"My family has owned and farmed (since 1932) properties, East Bonville, Lyons Road, Boambee Bridge area, Lamberts Road, and only koala I've seen was in a tree in my backyard in 18th Avenue in 1988, apart from 1 in Botanic Gardens". (Female, 58, Sawtell)

A consensus also emerged with regard to the Koala population of Korora. Older respondents observed a decline in the Koala population beginning in the 1980s:

"We had many koalas on our 5 acres 18 years ago and they went fairly quickly once the western side of Old Coast Road was opened up to more houses, particularly on the southern end of the road". (Female, 49, Korora)

"Koala bears were always round us living in Korora and then from approximately 1987 they disappeared". (Male, 65, Korora)

"I have six acres of trees with plenty of feed trees, also I adjoin the Orara East State Forest, but the amount of koalas seem to have declined over the last 20 years". (Male, 60, Korora)

Consistent with the presence of Koala habitat in the town centre of Coffs Harbour in the first half of the twentieth century, many respondents noted that they had observed Koalas in the town centre, but that their numbers had declined in recent decades. One resident, aged 62, observed that she had seen Koalas "in dense scrub from Sewerage Treatment Works to rail bridge west of railway line in early 40s. In Victoria St. koalas up telegraph poles on several occasions (approx. 1940-1950)" (Female, 62, Coffs Harbour). Another notes that "There were lots in Korora area where I lived as a kid and also in Bray St. area (1960 to 1970) in old Coffs Motel grounds – there were 32 acres there onto Bray St. and lots of suitable trees for koala round Coffs Creek tributary. Most of this land has been cleared" (Female, 45, Bayldon). Another respondent supports this claim: "Koalas were not infrequent in timber along Coffs Creek adjacent to Zara Pl. in early 70s. They appear to have gradually disappeared with development" (Male, 59, Coffs Harbour). Other comments point to the presence of Koalas in suburban areas:

"During 1960-65 we lived near Halls Road and saw koalas often, high in the trees". (Female, 70, Coffs Harbour)

"Soon after we moved to our present address about 1974 a koala was right near our front door" (Female, 61, Coffs Harbour)

"As a child living in Pitt Square Coffs Harbour I remember seeing koalas a few times in trees around our home – don't now". (Female, 35, Boambee)

Although these comments testify to the declining presence of Koalas in Coffs Harbour's urban and suburban areas, it is evident that small, semi-isolated populations persisted in these areas into the 1980s. *The Coffs Harbour Advocate* reported that, in late 1980, a Koala was found in the 'Target' store located in the town's central business district (*CHA* 1980d). However, it would be misleading to claim that such incidents reflected a healthy and stable population. A week after the 'Target' piece was published, the paper carried a front-page article entitled "Disappearing Haven", accompanied by a photograph of a Koala in a tree, which reported one Korora resident calling for a tree preservation order for Coffs Harbour Shire (*CHA* 1980e). It quoted the resident as stating that, without this, "Coffs Harbour was in danger of looking like one of the treeless Sydney suburbs" due to routine clearing carried out "without a thought for the local wildlife" (*CHA* 1980e). This view is consistent with the comments of many respondents to the Survey, which indicate a decline in the Koala population of the Coffs Harbour township beginning in the 1980s. As one respondent noted, "We did see koalas when we first lived here, 10 years ago, but not so much the past 2 years" (Female, 41, Coffs Harbour). The 1980s and 1990s were marked by particularly intense local interest in Koala conservation. Media coverage of the issue went into reached a high point,

with a total of 38 articles, including 3 editorials, in the *Advocate* in 1990 alone (*CHA* 1990a-al). This raises the possibility of a 'feedback loop' between local media and respondents' comments to the Survey, having taken place in 1990. However, while it is undeniable that intense media coverage of the issue heightened residents' awareness of Koalas in their area, the specificity of their comments – many detailing personal recollections and instances of habitat destruction – indicates that it is highly unlikely that media scrutiny *determined* these perceptions.

The articles can be broadly divided into four categories. Firstly, a number of articles publicise Local and State Government conservation and research initiatives related, but not limited to, the Coffs Harbour LGA (*CHA* 1990b,c,k,v,ac,ad,ag,ak). A smaller number of public interest pieces report on Koalas more generally, i.e., without recourse to local debates (*CHA* 1990c,g). One example of this is a report on the findings of a conference on Koala conservation held in Lismore (*CHA* 1990e). A third group of articles falls into the 'community interest' category, distinguished by an amusing tone and/or a presentation of Koalas as cute and cuddly (*CHA* 1990w,aj,al). These articles are generally removed from political debates and are accompanied by large photographs. The final group of articles is the largest, and focuses on local debates concerning Koala conservation and the efforts of local activist groups to place Koalas on the political agenda (*CHA* 1990a,d,f,h-j,l-u,x-z,aa,ab,ae,af,ah,ai).

The articles offer a wealth of information with regard to both the persistence, and rapid eradication, of Koala habitat in the Coffs Harbour LGA, in addition to the level of community interest in the issue. Key focal points include the urbanisation of North Bonville, which involved the illegal logging of live trees (*CHA* 1990u,x,z,aa); clearing at Bonville West for a golf course (*CHA* 1990t); and attempts to secure a Koala reserve at Roberts Hill (*CHA* 1990o). In particular, debates surrounding the development of Bonville made Koala conservation "a key by-election issue" for local government (*CHA* 1990ah). Landowners and developers labelled attempts by local conservation groups to secure land from development by extending tree preservation orders "play[ing] the koala card" (*CHA* 1990m). Protests reached their apogee in late 1990 when conservation groups presented Mayor Bernie Malouf with a dead Koala allegedly found floating in Pine Creek at Bonville (*CHA* 1990r). Malouf was already infamous for his widely publicised stance that private land should constitute an exception from measures intended to conserve Koala habitat (*CHA* 1988).

By 1990, as these articles demonstrate, the Koala came to symbolise the helplessness of Australia's native fauna in the face of relentless land clearing. It is important to note that this is merely a change in perception, as clearing at this juncture was no worse, qualitatively speaking, than that of a century earlier. In addition, the sudden increase in interest in Koalas could be mistaken for a rapid increase in the local Koala population. However, this too is misleading. Rather, we can draw two important conclusions from the local media coverage. Firstly, taken collectively, the articles either assume a neutral stance on the subject of Koala decline, or actively point to a decline and attribute this to multiple human-driven threats, primarily unchecked development. Although dissenting voices are present within some articles, not a single article attempts to deny the issue. This indicates that not only had the issue attained a critical political threshold, but that the Koala population was widespread enough throughout the LGA and that there were enough visible individual Koalas for local residents to form an opinion on the basis of personal experience. This does not necessarily point to a high population; rather, it indicates that Koalas were sufficiently present to be noticeable. Indeed, it could indicate a low-density population that was becoming progressively more visible as their habitat was fragmented by roads and clearing.

Secondly, the articles allow us to track the emergence of a preventative, and more holistic, approach to Koala conservation. Whereas development was portrayed in a positive light over the course of the 1960s to mid-1980s, it is considered critical by 1990 to control it utilising legal instruments such as tree preservation orders. Development is by this point perceived as the primary factor underlying all other threats to Koalas in the area. With this in mind, one resident writes, well-intentioned plans to build a hospital for sick and injured Koalas in Coffs Harbour (*CHA* 1990i) ultimately miss the mark:

Just as preventative medicine is about maintaining good health before sickness occurs, the health of koalas needs to be considered in terms of what are the causes of the major health risks to them. Looking beyond the immediate symptoms of having sick and injured koalas, it becomes necessary to ask what are the reasons behind such 'health problems'. Is it not the destruction of habitat through large scale clearing, encroaching suburban development with its accompanying threats of domestic pets and human traffic? The idea of rescuing sick and injured animals is a noble one, but somewhat naïve and short-sighted if it is not done in conjunction with a commitment to safeguarding viable areas of koala habitat. (*CHA* 1990f)

A few weeks later, the Editor of the *Advocate* espouses the same view in his editorial. After acknowledging that "Coffs Harbour needs to recognise the responsibility it owes its koalas and other wildlife", he argues that "While the idea of a hospital and wildlife refuge is admirable, it is in itself not a solution to the continuing conflicts between development and wildlife. The very fact that a hospital is needed suggests that strategies must be developed which will keep the animals out of the hospital." In his view, these strategies must include "a thorough audit" of the Koala population of the area, responsible development policies, and controls on domestic cats and dogs. He concludes: "There seems little point in patching up koalas at a hospital only to release them back into an environment in which they cannot survive" (*CHA* 1990j).

The evolution of this perspective led, in the early 2000s, to important measures designed to safeguard the existing Koala population from further threats. These measures included the ratification of the CKPoM by Coffs Harbour City Council in 1999 and the State Government in 2000 (*CHA* 2000d), the creation of the Bonville wildlife overpass (*CHA* 2000a), community initiatives such as planting Koala food trees in Coffs Harbour (*CHA* 2000h), and attempts to protect Pine Creek State Forest, in the south-western sector of the LGA (*CHA* 2000b,c). More broadly speaking, the focus of conservation action and population interest lay in the south-east sector of the LGA, which is consistent with ecological studies that identify this as the predominant location of Koalas in the LGA. While the focus of local media coverage lay, in the early 2000s, on a number of specific locations of contention, we can identify from the changing site-specific arguments that Koala populations remained present and still faced challenges. Foremost among these challenges was loss of habitat, with areas marked as primary Koala habitat in the CKPoM cleared in late 2000 (*CHA* 2000e,f,g).

Loss of habitat remains a key issue in 2015. This issue has two components. One is the 'legacy effect' of habitat loss and fragmentation that took place over previous decades, wherein individual Koalas which stayed in their home ranges will die and will not be replaced. The second is the additional loss of patches of habitat as individual developments proceed on vacant lots within housing estates, making these estates denser and less hospitable to Koalas. With increasing housing density, there is a corresponding increase in vehicle traffic and the presence of domestic dogs. Thus the Koala population becomes increasingly caught in a 'pincer movement' of decreasing habitat

34

Proc. Linn. Soc. N.S.W., 138, 2016

Fig. 28. View of Koala Place, a suburban cul-de-sac located near Boambee Creek, July 2014. Photograph by Dan Lunney.

and increasing threats. As photographs taken in January 2014 (Appendix 1) and July 2014 show, the once rural and forested landscape is now modern and suburban, with traces of Koala habitat remaining alongside creeks and on ridges. This is evident in Fig. 28, which depicts the ironically-named Koala Place, a suburban cul-de-sac located near Boambee Creek, where Koalas are still occasionally heard by residents.

CONCLUSION

As we have acknowledged, it is difficult to ascertain the precise pattern of change in the size of the Koala population of Coffs Harbour, largely due to the scarcity of relevant historical sources. In view of the absence of a population baseline until the late twentieth century, when Koala habitat was identified and mapped in the CKPoM, we must rely on alternative historical sources to trace changes to the Koala population. A general pattern can be drawn of an historical process stretching from the European settlement of Coffs Harbour to 2000 and inclusive of the broader Aboriginal pre-history of the area. The presence of the Koala in the languages, cultural practices and mythologies of the Gumbaynggir peoples indicates that Koalas had been present throughout the broader region prior to European settlement. The

relatively late arrival of European settlers in the Coffs Harbour area meant that Koala habitat was untouched until the early 1880s, when the local timber industry experienced a rapid boom. After the area was opened to selection in 1886, clearing became more extensive and timber rapidly became the area's primary export. In ecological terms, the timber industry presented at this time the dominant factor in the fragmentation and diminution of Koala habitat.

Unlike in Bega, on the far south-coast of NSW, where there were bear-skinning factories (Lunney and Leary 1988), we may conclude that the trade in marsupial skins and furs did not constitute a significant threat to the Koala population of Coffs Harbour in the late nineteenth and early twentieth centuries. While the trade was active in the broader region for a number of decades, it is evident that the industry was not prominent in Coffs Harbour itself. Drawing on the existing records, we may conclude that Koalas were present in Coffs Harbour in the early period of European settlement, but arguably never in high numbers.

In contrast to the fur trade, the timber industry continued to present the most significant threat to the Koala population of the area, expanding with the advent of new technology and the opening of the Coffs Harbour Jetty. As shown in historical photographs, extensive and unchecked vegetation clearing, logging, and ringbarking transformed the

Proc. Linn. Soc. N.S.W., 138, 2016

35

previously forested landscape into an agricultural landscape surrounded by forested hills. In particular, the pattern of clearing left the vegetation bordering creeks in the Coffs Harbour township, such as Coffs Creek, relatively intact due to the risk of floods. This facilitated the persistence of Koala habitat in these areas well into the 1980s, with some habitat remaining today. In contrast, the lands well above sea level, such as the Orara farms mentioned in this paper, were comprehensively cleared for farming, with little vegetation left along river edges. The forest on the slopes was left relatively intact, though it was progressively diminished throughout the twentieth century.

Our historical analysis has allowed us to determine the extent to which other potential threats affected the Koala population of the area. In particular, it has allowed us to eliminate fire as a major threat to Koala populations in the area over the twentieth century. Mapping the distribution of fires shows that they were scattered and sufficiently infrequent, thereby failing to comprise a major threat. Though fires would have killed Koalas inhabiting the areas where the fires occurred, their long-term impact on the presence of Koala populations can be considered negligible due to the rapid rate of recovery of the forest as Koala habitat. As the surrounding unburnt forests were extensive in area, they would have provided a crucial source population for the rapid recolonisation of the burnt areas as these areas recovered. Historical sources have also enabled us to qualify the extent to which human population growth has constituted a threat to Koalas over time. Following a marked decline in the human population during the 1920s and Great Depression, population growth only began to constitute a threat after 1945 as the population steadily increased. However, given the slow pace of this demographic shift, we can assume that population growth did not constitute a major threat to the Coffs Harbour Koala population prior to the 1960s, when the area experienced a boom. Since then, growth has continued unabated, with the rate of human population expansion in the area exceeding the growth rate of NSW as a whole by 2000.

These population trends provide a context for the increasing degradation and fragmentation of Koala habitat in the area. As a result of the increasing demand, beginning in the mid-1960s, for housing and associated infrastructure to meet the needs of the growing human population, land which is now recognised to have contained Koala habitat was progressively cleared and converted into an urban landscape containing habitat fragments. The glaring transformation of a rural-forest landscape in 1964 to the suburban estates of today, as shown in aerial photography, supports the general thesis of an incremental loss of habitat long before there was ever a scientific definition of Koala habitat. This process was accompanied by a number of associated threats, most significantly the growing presence of vehicle traffic – leading to roadkill – and domestic dogs.

Taken collectively, the evidence allows us to draw a number of conclusions: that the Koala population of Coffs Harbour was widespread but never abundant, that habitat loss has been relentless since European settlement, and that the fur trade in Koala skins was not extensive in the late nineteenth and early twentieth centuries. The transformation of a rural-forest to an urban landscape, particularly in the south-east of Coffs Harbour, over the past four decades is the most recent stage in the incremental loss of habitat since European settlement. Consequently, the conclusion can be drawn that the Koala population had been reduced from its pre-European size by 1990. It is important to recognise that these trends are specific to the Koala population of Coffs Harbour. It is our contention that, in order to fully understand the threats facing Koala populations, we must examine these populations within their local context. The relative significance of different threats, particularly habitat loss, varies among localities (McAlpine et al. 2006, 2008). Our ecological history shows that threats which have exerted a significant effect on Koala populations of other areas, such as the fur trade, are of lesser importance in the Coffs Harbour LGA. By contrast, the single most significant factor in the historical decline of the Koala population in this area has been a continual process of habitat loss and fragmentation, compounded in the late twentieth century by extensive development to accommodate an increasing human population.

As the CKPoM and current ecological research shows, habitat loss and fragmentation continue to present a threat to Koala populations in the area. However, the remaining patches of native forest will continue to attract Koalas because they are core Koala habitat. In our view, the continuing presence of Koalas in suburban areas is a misleading indicator of the survival of the population as a whole, as these individual Koalas are likely to have emigrated from forest elsewhere, such as Bongil Bongil National Park in the south-east of the Coffs Harbour LGA. It is the lethal impact of vehicles and dogs in the exposed stretches between the habitat fragments that will arguably cause relentless loss within the Koala population in the south-east of the Coffs Harbour LGA. A detailed radio-tracking and demographic study of the metapopulation is needed to determine

36

Proc. Linn. Soc. N.S.W., 138, 2016

in what locations the local Koala populations are persisting, declining and migrating, with measures of health, fertility and mortality. Such a study would include Bongil Bongil National Park in conjunction with the urban and peri-urban areas of south-east Coffs Harbour LGA.

The catalogue of new or rising threats, such as roadkill, dogs, and disease (Lunney et al. 2015), in addition to the future threat of climate change, compounds the long-term threatening processes of habitat loss and fragmentation that we have traced in this paper. Studies of Koala populations in other areas, using different methods, have allowed us to gauge the relative significance of these threats for Coffs Harbour. Specifically, a detailed radio-tracking study of Koalas in Port Stephens determined that dogs were a major, but unseen, killer of Koalas (Lunney et al. 2007), whereas fire and roadkill were more conspicuous (Matthews et al. 2007; Rhodes et al. 2014) but not necessarily as significant in the Coffs Harbour context. Population and modelling studies in the Eden region of south-east NSW, Gunnedah in north-west NSW and in Queensland help to determine the impact of climate change on the Koala populations of these areas (Lunney et al. 2012, 2014; Adams-Hoskings et al. 2011, 2014), demonstrating that it presents a widespread and insidious threat, and one that will invariably affect the Koala population of Coffs Harbour.

Proposed actions for reversing the decline of the Koala in NSW are presented in the NSW Koala Recovery Plan (DECC 2008) and the National Strategy for the Conservation and Management of the Koala 2009-14 (Commonwealth of Australia 2009). Despite the apparent clarity of these strategies, ambiguities in our contemporary understanding of the Koala may complicate attempts to reverse their decline. The Senate Committee's report on its 2011 enquiry into the Koala expressed surprise at what it called the "complexity of this multifaceted issue" (Commonwealth of Australia 2011, xv). When the Committee looked into the Koala question, it was inundated with submissions pointing to current problems, but there was not a series of ecological histories of Koala populations to assist in interpreting changes. Nor was the need for an ecological history of the Koala identified in the 19 recommendations of the Senate Committee for action (Shumway et al. 2015). In 2012, the Commonwealth Government listed the Koala as a threatened species in ACT, NSW and Queensland, thereby confirming the decline that had become obvious in many locations, especially along coastal NSW, such as Coffs Harbour, and nearby Iluka, where the population became effectively extinct in

the late twentieth century (Lunney et al. 2002).

In our view, if we limit our focus to contemporary issues facing existing Koala populations, we are likely to overlook the causes of long-term change and to mismanage what remains of our faunal heritage. The threatened species status of Koalas under both Commonwealth and State legislation and the ratification of the 1999 CKPoM for Coffs Harbour all represent moves in the right direction for Koala conservation, but given our interpretation of long-term change, these policy documents alone will not stem the continual contraction of the Koala population of Coffs Harbour. While ecological history is indispensable for deepening our understanding of long-term change, complementary studies are needed to pinpoint the impact of specific threats. There is a pressing need in Coffs Harbour for a population study that moves beyond the identification of shifts in population distribution and habitat mapping, to examining other population attributes such as rates of breeding and mortality, and the dynamics of Koala immigration and emigration. This ecological information is critical in identifying long-term patterns, interpreting current changes to a population profile, and developing strategies to manage threats. It is the interaction of the historical and ecological approaches, as demonstrated in this study, which will allow us to most effectively understand and manage Koala populations of specific regions.

ACKNOWLEDGEMENTS

We thank Dan Lunney's co-authors on another project, Bradley Law and Catherine Rummery, for their unpublished research on the Australian fur trade, and Martin Predavec for his helpful comments. We also thank John Turbill and Chris Moon for their comments and insights over the years, and Nigel Cotsell and Rachel Binskin of Coffs Harbour City Council for their assistance with material. We also thank two anonymous referees for their comments.

REFERENCES

Adams-Hosking, C., Grantham, H. S., Rhodes, J. R., McAlpine, C., & Moss, P. T. (2011). Modelling climate-change-induced shifts in the distribution of the koala. *Wildlife Research* **38**(2): 122-130.
Adams-Hosking, C., McAlpine, C. A., Rhodes, J. R. Moss, P. T. and Grantham H. S. (2014). Prioritizing regions to conserve a specialist folivore: considering probability of occurrence, food resources, and climate change. *Conservation Letters* [doi: 10.1111/conl.12125].

Proc. Linn. Soc. N.S.W., 138, 2016

37

KOALAS IN COFFS HARBOUR

Argus, The. (1880). The Australian Fur Trade. (9 December): 52.

Australian Bureau of Statistics [ABS]. (1991). Census Counts for Small Areas: New South Wales. 1991 Census of Population and Housing. (Canberra, A.C.T.: Commonwealth Government Printer): 20.

Australian Bureau of Statistics [ABS]. (2001). Basic Community Profile. Available at: <http://www.censusdata.abs.gov.au/census_services/getproduct/census/2001/communityprofile/LGA11800> [Accessed 11 September 2014].

Australian Government Department of the Environment. (2011). Advice to the Minister for Sustainability, Environment, Water, Population and Communities from the Threatened Species Scientific Committee (the Committee) on Amendment to the list of Threatened Species under the Environment Protection and Biodiversity Conservation Act 1999 (EPBC Act). Available at: <http://www.environment.gov.au/biodiversity/threatened/species/pubs/197-listing-advice.pdf> [Accessed 14 February 2015].

Bacon, V. F. (1926). 'Guide Booklet to Coffs Harbour and District'. (Coffs Harbour, N.S.W.: Coffs Harbour Chamber of Commerce).

Bellingham, S.R. (1900). Native Bear. *Geo. Wilcox & Co.'s Review.* (5 September): 98.

Clarence and Richmond Examiner, The. (1886a). Young native bear for sale. (5 January): 1.

Clarence and Richmond Examiner, The. (1886b). Young native bear for sale. (9 January): 1.

Clarence and Richmond Examiner, The. (1889a). Commercial Report. (3 August): 4.

Clarence and Richmond Examiner, The. (1889b). The Doom of the Kangaroo. (3 August): 6.

Clarence and Richmond Examiner, The. (1889c). Commercial Report. (10 August): 4.

Clarence and Richmond Examiner, The. (1889d). Close Season for the Kangaroo. (13 August): 4.

Clarence and Richmond Examiner, The. (1889e). Commercial Report. (17 August): 4.

Clarence and Richmond Examiner, The. (1889f). Commercial Report. (24 August): 4.

Clarence and Richmond Examiner, The. (1889g). Commercial Report. (31 August): 4.

Clarence and Richmond Examiner, The. (1889h). Commercial Report. (14 September): 4.

Clarence and Richmond Examiner, The. (1889i). Commercial Report. (28 September): 4.

Clarence and Richmond Examiner, The. (1889j). Commercial Report. (1 October): 2.

Clarence and Richmond Examiner, The. (1889k). Commercial Report. (5 October): 4.

Clarence and Richmond Examiner, The. (1893). Kangaroo Farms. (11 March): 5.

Clarence and Richmond Examiner, The. (1894). Coff's Harbour. (6 January): 3.

Clarence and Richmond Examiner, The. (1895). Rambling Notes: From the Bellinger to Grafton. (13 August): 8.

Clarence and Richmond Examiner, The. (1896). Coffs Harbour. (4 January): 5.

Clarence and Richmond Examiner, The. (1898). Coff's Harbour Trade. (22 January): 4.

Clarence and Richmond Examiner, The. (1899). Coff's Harbour. (14 January): 2.

Clarence and Richmond Examiner, The. (1902). Coffs Harbour Exports and Imports. (14 January): 4.

Clarence and Richmond Examiner, The. (1905). From Grafton to Coff's Harbour. (25 November): 3.

Clarence and Richmond Examiner, The. (1907). Exports from the North Coast. (12 January): 11.

Clarence and Richmond Examiner, The. (1908a). Exports from the North Coast. (26 September): 10.

Clarence and Richmond Examiner, The. (1908b). Exports from the North Coast. (3 October): 10.

Clarence and Richmond Examiner, The. (1909a). Footing it to Queensland. From "Port" to Coff's Harbour. (1 May): 3.

Clarence and Richmond Examiner, The. (1909b). A Trip to the Coast. Coff's Harbour and the Timber Industry. (17 June): 2.

Clarence and Richmond Examiner and New England Advertiser, The. (1886). The Orara Reserves. (24 July): 3.

Clarence and Richmond Examiner and New England Advertiser, The. (1889). The Skin Trade. (13 July): 4.

Coffs Harbour Advocate, The. (1910a). Advertisement. Goldsborough, Mort & Co. (4 March): 6.

Coffs Harbour Advocate, The. (1910b). Advertisement. Pastoral Finance Association Limited. (7 October): 1.

Coffs Harbour Advocate, The. (1920). Flying Fox Pest. (28 March): 1.

Coffs Harbour Advocate, The. (1930). Opossums and Bananas. (8 July): 3.

Coffs Harbour Advocate, The. (1940). Wallabies a Pest. (13 August): 3.

Coffs Harbour Advocate, The. (1950). Interesting Reminiscences of Former District Resident. (17 October): 6.

Coffs Harbour Advocate, The. (1960a). Much of our Rare and Beautiful Fauna is Being Lost. (20 January): 1.

Coffs Harbour Advocate, The. (1960b). Need for more Fauna Protection Reserves. (25 March): 1.

Coffs Harbour Advocate, The. (1960c). Fauna export ban. (7 April): 2.

Coffs Harbour Advocate, The. (1960d). Wanton destruction of fauna on North Coast. (21 April): 10.

Coffs Harbour Advocate, The. (1960e). Action to stop shooting of protected fauna. (2 August): 1.

Coffs Harbour Advocate, The. (1970a). Clearing the jungle. (13 February): 1.

Coffs Harbour Advocate, The. (1970b). Letter: Protect the Koala. (13 May): 2.

Coffs Harbour Advocate, The. (1970c). Hillside Being Cleared for New Crown Land Subdivision. (3 August): 3.

Coffs Harbour Advocate, The. (1970d). Letter: Protection of Wildlife. (14 September): 2.

Coffs Harbour Advocate, The. (1970e). Move to Protect the Red Kangaroo. (14 September): 10.

38

Proc. Linn. Soc. N.S.W., 138, 2016

Coffs Harbour Advocate, The. (1980a). Coffs Creek urban development plan. (19 August): 3.

Coffs Harbour Advocate, The. (1980b). Massive land scheme gets green light. (29 August): 1.

Coffs Harbour Advocate, The. (1980c). MP urges new council: Re-route Highway bypass. (9 September): 1.

Coffs Harbour Advocate, The. (1980d). Untitled. (10 October): 2.

Coffs Harbour Advocate, The. (1980e). Disappearing Haven. (17 October): 1.

Coffs Harbour Advocate, The. (1988). Public land only for koalas, says Malouf. (3 December): 2.

Coffs Harbour Advocate, The. (1990a). Letter: Koala times. (12 January): 4.

Coffs Harbour Advocate, The. (1990b). Studies into koala, potoroo habitats. (1 February): 3.

Coffs Harbour Advocate, The. (1990c). Project to identify koala habitat. (3 March): 12.

Coffs Harbour Advocate, The. (1990d). Koalas topic of talk. (4 April): 12.

Coffs Harbour Advocate, The. (1990e). Cars, dogs 'main killers' of koalas. (9 May): 6.

Coffs Harbour Advocate, The. (1990f). Letter: Koala hospital. (12 June): 4.

Coffs Harbour Advocate, The. (1990g). Koalas under threat. (27 July): 19.

Coffs Harbour Advocate, The. (1990h). Hospital for koalas. (11 August): 1.

Coffs Harbour Advocate, The. (1990i). Caring for koalas. (11 August): 2.

Coffs Harbour Advocate, The. (1990j). Wildlife responsibility. (11 August): 4.

Coffs Harbour Advocate, The. (1990k). Coffs Koala Study. (16 August): 1.

Coffs Harbour Advocate, The. (1990l). Moke 'n' Matilda prove an armful. (16 August): 1.

Coffs Harbour Advocate, The. (1990m). Group wants tree order extended. (22 August): 2.

Coffs Harbour Advocate, The. (1990n). Tree preservation order 'out of order'. (28 August): 3.

Coffs Harbour Advocate, The. (1990o). Legal threat hangs over koala reserve. (1 September): 3.

Coffs Harbour Advocate, The. (1990p). 'Why the outrage over one decapitated koala?'. (6 September): 7.

Coffs Harbour Advocate, The. (1990q). Another view of Bonville. (7 September): 4.

Coffs Harbour Advocate, The. (1990r). Mayor clashes with Bonville protestors over dead koala. (11 September): 2.

Coffs Harbour Advocate, The. (1990s). Ranger explains role in caring for fauna. (12 September): 30.

Coffs Harbour Advocate, The. (1990t). Clearing for golf course is 'legal'. (14 September): 2.

Coffs Harbour Advocate, The. (1990u). Conservation group repeats inquiry call. (21 September): 2.

Coffs Harbour Advocate, The. (1990v). Survey to record koala numbers. (29 September): 3.

Coffs Harbour Advocate, The. (1990w). Giant koala family takes pride of place. (3 October): 1.

Coffs Harbour Advocate, The. (1990x). BBH denies breaking preservation order. (5 October): 2.

Coffs Harbour Advocate, The. (1990y). Jenny acts while others talk. (11 October): 1.

Coffs Harbour Advocate, The. (1990z). Koala claim. (12 October): 1.

Coffs Harbour Advocate, The. (1990aa). Koala habitat destroyed. (12 October): 2.

Coffs Harbour Advocate, The. (1990ab). Koalas need your help. (12 October): 4.

Coffs Harbour Advocate, The. (1990ac). Koalas on the move. (17 October): 2.

Coffs Harbour Advocate, The. (1990ad). Tree preservation order to protect sensitive land. (18 October): 5.

Coffs Harbour Advocate, The. (1990ae). Koala question for candidates. (20 October): 26.

Coffs Harbour Advocate, The. (1990af). Koala corpses. (25 October): 4.

Coffs Harbour Advocate, The. (1990ag). Koala database should be 'the best in State'. (26 October): 2.

Coffs Harbour Advocate, The. (1990ah). Bonville is a 'key' by-election issue. (26 October): 5.

Coffs Harbour Advocate, The. (1990ai). Letter: Koala protection. (17 November): 4.

Coffs Harbour Advocate, The. (1990aj). It's a hard life when you're not up a gum tree. (22 November): 3.

Coffs Harbour Advocate, The. (1990ak). Letter: Koala survey: council's role. (29 November): 4.

Coffs Harbour Advocate, The. (1990al). Animal circus to be a big hit over Christmas. (26 December): 4-7.

Coffs Harbour Advocate, The. (2000a). Wildlife overpass a first. (18 March): 3.

Coffs Harbour Advocate, The. (2000b). Unique plan to save 'our' koalas. (13 May): 2.

Coffs Harbour Advocate, The. (2000c). Keeping an eye on koalas. (24 June): 13.

Coffs Harbour Advocate, The. (2000d). Koala plan ensures 'long-term survival'. (30 June): 5.

Coffs Harbour Advocate, The. (2000e). Land clearing has residents in shock. (9 August): 3.

Coffs Harbour Advocate, The. (2000f). NPWS gets tough over koala land. (12 August): 19.

Coffs Harbour Advocate, The. (2000g). Ulitarra calls for controls. (16 August): 4.

Coffs Harbour Advocate, The. (2000h). With students' help, Blinky Bill won't go hungry. (25 October): 5.

Coltheart, L. (1997). 'Between Wind and Water: A history of the ports and coastal waterways of New South Wales'. (Sydney, N.S.W.: Hale & Iremonger).

Commonwealth of Australia. (2011). The Senate. Environment and Communications

Cumberland Argus and Fruitgrowers Advocate, The. (1908). Up North: A Local Man's Impressions. (27 June): 12.

DECC. (2008). 'Recovery Plan for the Koala *(Phascolarctos cinereus)*'. (Goulburn St, Sydney, N.S.W.: NSW Department of Environment and Climate Change).

KOALAS IN COFFS HARBOUR

Diarmid. (1903). The Shooting Industry of New South Wales. *Geo. Wilcox & Co.'s Review.* (7 January): 825-826.

Donlan, C. J., and Martin, P. S. (2004). Role of Ecological History in Invasive Species Management and Conservation. *Conservation Biology* **18**, 267-269.

England, G. (1976). 'The Coffs Harbour Story'. (Coffs Harbour, N.S.W.: The Central North Coast Newspaper Company).

Evening News, The. (1902). The Timber Industry. Mr Bennett's Visit. His Impressions of the Districts. (28 August): 4.

Flannery, T. (2001). 'The eternal frontier: an ecological history of North America and its peoples'. (Melbourne, VIC: Text Publishing).

Foster, D. (2000). Conservation lessons and challenges from ecological history. *Forest History Today* (Fall), 2-11.

Fuchs, V. (1957). 'The economics of the fur industry'. (New York: Columbia University Press).

Gordon, G., and Hrdina, F. (2005). Koala and Possum Populations in Queensland during the Harvest Period, 1906-1936. *Australian Zoologist* **33**(1): 69-99.

Gordon, G., Hrdina, F., and Patterson, R. (2006). Decline in the distribution of the Koala *Phascolarctos cinereus* in Queensland. *Australian Zoologist* **33**(3): 345-358.

Goulburn Evening Penny Post, The. (1908). The Northern Coast. (7 November): 4.

Grove, A., and Rackham, O. (2001). 'The Nature of Mediterranean Europe: An Ecological History'. (New Haven; London: Yale University Press).

Hobson, H.J. (1978). Discovery of the Bellinger. In 'Pioneering in the Bellinger Valley' (eds. N. Braithwaite and H. Beard) pp. 4-6 (Bellingen, N.S.W. : Bellinger Valley Historical Society).

Hodgkinson, C. (1845). 'Australia, from Port Macquarie to Moreton Bay; with descriptions of the natives, their manners and customs; the geology, natural productions, fertility, and resources of that region; first explored and surveyed by order of the colonial government'. (London: T. and W. Boone).

Hrdina, F., and Gordon, G. (2004). The Koala and Possum Trade in Queensland, 1906-1936. *Australian Zoologist* **32**(4): 543-584.

Jackson, S. T., and Hobbs, R. J. (2009). Ecological Restoration in the Light of Ecological History. *Science* **325**: 567-568.

Knott, T., Lunney, D., Coburn, D., and Callaghan, J. (1998). An ecological history of Koala habitat in Port Stephens Shire and the Lower Hunter on the Central Coast of New South Wales, 1801-1998. *Pacific Conservation Biology* **4**: 354-368.

Lassau, S., Ryan, B., Close, R., Moon, C., Geraghty, P., Coyle, A., and Pile, J. (2008). Home ranges and mortality of a roadside Koala *Phascolarctos cinereus* population at Bonville, New South Wales. In 'Too close for comfort: contentious issues in human-wildlife encounters' (eds D. Lunney, A. Munn, and

W. Meikle), pp. 127-136. (Mosman, N.S.W. : Royal Zoological Society of New South Wales).

Le Souef, A. S., and Burrell, H. (1926). 'The wild animals of Australasia'. (London: George G. Harrap and Co.).

Ling, J. (1999). Exploitation of fur seals and sea lions from Australian, New Zealand and adjacent subantarctic islands during the eighteenth, nineteenth and twentieth centuries. *Australian Zoologist* **31**(2): 323-350.

Lunney, D. (2001). Causes of the extinction of native mammals of the Western Division of New South Wales: an ecological interpretation of the nineteenth century historical record. *The Rangeland Journal* **23**: 44-70.

Lunney, D., Close, R., Bryant, J., Crowther, M.S., Shannon, I., Madden, K., and Ward, S. (2010). Campbelltown's koalas: their place in the natural history of Sydney. In 'The Natural History of Sydney' (eds. D. Lunney, P. Hutchings, and D. Hochuli) pp. 319-325 (Mosman, N.S.W. : Royal Zoological Society of New South Wales).

Lunney, D., Crowther, M.S., Wallis, I., Foley, W.J., Lemon, J., Wheeler, R., Madani, G., Orscheg, C., Griffith, J.E., Krockenberger, M., Retamales, M. and Stalenberg, E. (2012). Koalas and climate change: a case study on the Liverpool Plains, north-west NSW. In 'Wildlife and climate change: towards robust conservation strategies for Australian fauna' (eds. D. Lunney and P. Hutchings) pp 150-168. (Mosman, N.S.W. : Royal Zoological Society of New South Wales).

Lunney, D., Gresser, S., O'Neill, L.E., Matthews, A. and Rhodes, J. (2007). The impact of fire and dogs on koalas at Port Stephens, New South Wales, using population viability analysis. *Pacific Conservation Biology* **13**: 189-201.

Lunney, D., Law, B., and Rummery, C. (1997). An ecological interpretation of the historical decline of the Brush-tailed Rock-wallaby *Petrogale penicillata* in New South Wales. *Australian Mammalogy* **19**: 281-296.

Lunney, D., and Leary, T. (1988). The impact on native mammals of land-use changes and exotic species in the Bega district (New South Wales) since settlement. *Australian Journal of Ecology* **13**: 67-92.

Lunney, D., Matthews, A., Moon, C., and Ferrier, S. (2000). Incorporating habitat mapping into practical koala conservation on private lands. *Conservation Biology* **14**: 669-80.

Lunney, D., Matthews, A., Moon, C., and Turbill, J. (2002). Achieving fauna conservation on private land: reflections on a ten-year project. *Ecological Management and Restoration* **3**: 90-96.

Lunney, D., Moon, C., Matthews, A., and Turbill, J. (1999a). 'Coffs Harbour City Koala Plan of Management. Parts A & B'. (Hurstville, N.S.W. : NSW National Parks and Wildlife Service).

Lunney, D., Moon, C., Matthews, A., and Turbill, J. (1999b). 'Comments from Respondents to the

40

Proc. Linn. Soc. N.S.W., 138, 2016

1990/91 Postal Koala Survey of Coffs Harbour'.
(Hurstville, N.S.W.; Coffs Harbour, N.S.W.: NSW
National Parks and Wildlife Service and Coffs
Harbour City Council).

Lunney, D., O'Neill, L., Matthews, A. and Sherwin,
W.B. (2002). Modelling mammalian extinction
and forecasting recovery: koalas at Iluka (NSW,
Australia). *Biological Conservation* **106**: 101-13.

Lunney, D., Predavec, M., Miller, I., Shannon I., Fisher,
M., Moon, C., Matthews, A., Turbill, J., Rhodes, J.,
(2015).Interpreting patterns of population change
in koalas from long-term datasets in Coffs Harbour
on the north coast of New South Wales. *Australian
Mammalogy* http://dx.doi.org/10.1071/AM15019.

Lunney, D. Stalenberg, E., Santika, T. and Rhodes, J. R.
(2014). Extinction in Eden: identifying the role of
climate change in the decline of the koala in south-
eastern NSW. *Wildlife Research* 41:22-34.

Lunney, D., Urquhart, C.A., and Reed, P. (eds). (1990).
'Koala Summit. Managing koalas in NSW'.
(Hurstville, N.S.W. : NSW National Parks and
Wildlife Service).

Marshall, A. J. (1966). On the Disadvantages of
Wearing Fur. In A. J. Marshall (ed.), 'The Great
Extermination: A guide to Anglo-Australian
Cupidity, Wickedness and Waste' pp 9-42 (London:
Heinemann).

Matthews, A., Lunney, D., Gresser, S. and Mailz, W.
(2007). Tree use by koalas *Phascolarctos cinereus*
after fire in remnant coastal forest. *Wildlife Research*
34: 84-93.

Mayers, M. (1987). Voice of Time [sound recording]: Oral
history interview with Mabel Myers. Interviewed
by Sheridah Melvin (6 February). (Coffs Harbour,
N.S.W.: Coffs Harbour City Library).

McAlpine, C. A., Bowen, M. E., Callaghan, J. G., Lunney,
D., Rhodes, J. R., Mitchell, D. L., Possingham,
H. P. (2006). Testing alternative models for the
conservation of koalas in fragmented rural urban
landscapes. *Austral Ecology* 31(4): 529-529.

McAlpine, C.A., Rhodes, J.R., Bowen, M.E., Lunney, D.,
Callaghan, J.G., Mitchell, D.L. and Possingham, H.P.
(2008). Can multiscale models of species' distribution
be generalized from region to region? A case study of
the koala. *Journal of Applied Ecology* **45**: 558-567.

McFarlane, D. (1934a). Aboriginals: Mode of Living,
Clarence River Tribes. *Grafton Daily Examiner* (9
April), p. 7.

McFarlane, D. (1934b). Aboriginal History: Corroborees
and Bora Ceremonies, Methods of Hunting and
Foraging, Opossum a Favourite Dish. *Grafton Daily
Examiner* (19 May), p. 12.

Menkhorst, P. (2008). Hunted, marooned, re-introduced,
contracepted: a history of Koala management in
Victoria. In 'Too close for comfort: contentious issues
in human-wildlife encounters' (eds D. Lunney, A.
Munn and W. Meikle), pp. 73-92. (Mosman, N.S.W. :
Royal Zoological Society of New South Wales).

Moon, C. (1995). W.I.R.E.S. koala statistics tell a story,
pp. 49-54. (Proceedings of a conference on the

status of koalas in 1995, 4th Nat. Carers Conf., AKF
Brisbane),

Moyal, A. M. (2008). 'Koala: A Historical Biography'.
(Collingwood, VIC: CSIRO Publishing).

N.S.W. Government Gazette. (1861). No. 72. (24
December): 2765.

N.S.W. Government. (1903). Act No. 18: Native Animals
Protection. An Act to protect native animals, and to
amend the Birds Protection Act, 1901. (5 December):
74-76. Accessible at: <http://www.austlii.edu.au/au/
legis/nsw/num_act/napa1903n18300.pdf> (Accessed
7 June 2014).

North Western Courier, The. (1939). Koala on Road. (4
December): 4.

Parris, H.S. (1948). Koalas on the lower Goulburn.
Victorian Naturalist 64:192-193.

Pegum, M., and Pegum, S. (2010). 'Crossing the Bar:
A History of the Urunga Pilot Station'. (Urunga,
N.S.W.: Pilot House).

Petersen, M. (1914). 'The Fur Traders and Fur-Bearing
Animals'. (Buffalo, N.Y.: The Hammond Press).

Poland, H. (1892). 'Fur-Bearing Animals in Nature and in
Commerce'. (London: Gurney & Jackson).

Predavec, M., Lunney, D., Hope, B., Stalenberg, E,
Shannon, I., Crowther, M.S. and Miller I. (in
press). The contribution of community wisdom to
conservation biology? *Conservation Biology*.

Queenslander, The. (1905). Australian Fur Skin Trade. (11
February): 4.

Raleigh Sun, The. (1900). Coff's Harbour. (13 April): 4.

Reed, P., and Lunney, D. (1990). Habitat loss: the key
problem for the long-term survival of koalas in New
South Wales. In Daniel Lunney, Chris Ann Urquhart
and Philip Reed, 'Koala Summit: Managing koalas
in New South Wales' pp. 9-31 (Hurstville, N.S.W.:
National Parks and Wildlife Service).

Reed, P., Lunney, D., and Walker, P. (1990). A 1986-
1987 survey of the Koala *Phascolarctos cinereus*
(Goldfuss) in New South Wales and an ecological
interpretation of its distribution. In A.K. Lee, K.A.
Handasyde and G.D. Sanson (eds.), 'Biology of the
Koala' (Sydney, N.S.W.: Surrey Beatty & Sons): 55-
74.

References Committee. 'The koala—saving our national
icon'. (Parliament House, Canberra, A.C.T.: Senate
Printing Unit).

Rhodes, J.R., Lunney, D., Callaghan, J., McAlpine, C.A.
(2014). A few large roads or many small ones?
How to accommodate growth in vehicle numbers
to minimise impacts on wildlife. PLoS ONE 9(3):
e91093. doi: 10.1371/journal.pone.0091093.

Richards, M. (1996). 'North Coast run: men and ships of
the NSW North Coast'. (Wahroonga, N.S.W.: Turton
& Armstrong).

RL Newman and Partners Pty Ltd. (1996). 'A brief
history of forestry management at Pine Creek for the
Forestry Commission of NSW'. Report No. 1112
to SFNSW Coffs Harbour Region. (Canberra: RL
Newan and Partners).

Proc. Linn. Soc. N.S.W., 138, 2016

41

Rodwell, M. (2011). Coffs Harbour's European History. In Nan Cowling (ed.), 'Coffs Harbour Time Capsule Book: 1847-2011', vol. 3 (Coffs Harbour, N.S.W.: Published by Nan Cowling): 27.

Rudder, E. (1899). Native Names of Places --- from Aborigines on the Orara River, *Science of Man* 2 (21 September): 144.

Ryan, J. (1964). The Bear and the Water: A Study in Mythological Etymology, *Folklore* 75 (Winter): 260-268.

Ryan, J. (1988). 'The Land of Ulitarra: Early Records of the Aborigines of the Mid-North Coast of New South Wales'. (Lismore, N.S.W.: N.S.W. Department of Education).

Secomb, M. (1986). 'Red Gold to Green Grass: The Early History of the Upper Orara Valley'. (Upper Orara-Karangi Centenary Association: Coffs Harbour).

Shumway, N., Lunney, D., Seabrook, L., and McAlpine, C. (2015). Saving our national icon: An ecological analysis of the 2011 Australian Senate inquiry into status of the koala. *Environmental Science and Policy*. http://dx.doi.org/10.1016/j.envsci.2015.07.024.

Smith, A. (2004). Koala conservation and habitat requirements in a timber production forest in north-east New South Wales. In Daniel Lunney (ed.), 'Conservation of Australia's Forest Fauna' pp. 591-611 (Mosman, N.S.W.: The Royal Zoological Society of New South Wales).

Sydney Morning Herald, The. (1906). Timber and Land Settlement. An Open Invitation. (28 September): 4.

Sydney Morning Herald, The. (1910). Australian Fauna. To the Editor of the Herald. (30 August): 4.

Sydney Morning Herald, The. (1936a). Bushfires. Severe Damage. In Hardwood Belt. (2 November): 10.

Sydney Morning Herald, The. (1936b). Anxiety for Orara. (4 December): 19.

Sydney Morning Herald, The. (1936c). Bushfires. Township Saved. (5 December): 18.

Sydney Morning Herald, The. (1951a). Bushfires Destroy Valuable Timber. (2 October): 3.

Sydney Morning Herald, The. (1951b). Bushfires Devastate 64 State Forests. (26 October): 1.

Sydney Morning Herald, The. (1951c). Fierce Bushfires Ravage North Coast. (27 October): 3.

Sydney Morning Herald, The. (1953). Coff's Harbour: Bushfire Danger Eases. (19 November): 5.

Sydney Stock and Station Journal, The. (1904). Winchcombe, Carson & Co. Ltd. Report. (12 July): 2.

Sydney Stock and Station Journal, The. (1905). Winchcombe, Carson & Co. Ltd. Report. (18 April): 2.

Thomas, L. (2013). 'Aboriginal history of the Coffs Harbour region'. (Coffs Harbour, N.S.W.: Coffs Harbour City Library).

Tindale, N. (1940). Distribution of Australian Aboriginal Tribes. *Transactions of the Royal Society of South Australia* 64:1 (26 July): 140-231.

Troughton, E. (1948). 'Furred animals of Australia'. (Sydney, N.S.W.: Angus and Robertson).

Turbill, J. (2014). Pers. comm. (11 June).

Tyrrell, J. (1953). 'Australian Aboriginal Place-Names and their Meanings'. (Sydney, N.S.W.: Tyrrell's).

Vermeij, G. (1987). 'Evolution and Escalation: An Ecological History of Life'. (Princeton: Princeton University Press).

Yeates, N. (1990). 'Coffs Harbour, Vol. I: Pre-1880 to 1945'. (Coffs Harbour, N.S.W.: Bananacoast Printers for Coffs Harbour City Council).

Yeates, N. (1993). 'Coffs Harbour, Vol. II: 1946 to 1964'. (Coffs Harbour, N.S.W.: Bananacoast Printers for Coffs Harbour City Council).

Zu Ermgassen, P., Spalding, M., Blake, B., Coen, L., Dumbauld, B., Geiger, S., Grabowski, J., Grizzle, R., Luckenbach, M., McGraw, K., Rodney, W., Ruesink, J., Powers, S., and Brumbaugh, R. (2012). Historical ecology with real numbers: past and present extent and biomass of an imperilled estuarine habitat. *Proc. R. Soc. B* 279: 3393-3400.

42

Proc. Linn. Soc. N.S.W., 138, 2016

APPENDIX 1

Fig. 1. This 2009 high-resolution aerial photo, in the ADS40 series, shows the present-day location of both Hoschke's and McLeod's farms. At the centre of the photo is a track that crosses the Orara River. The River runs vertically and centrally through most of the photo, then turns left near the top of the photo. The historical photo of Hoschke's farm (Fig. 7) was taken from just below the main cluster of buildings and to the right of the centre line. McLeod's farm is on the right hand side of the River, and occupies much of the centre of the right of the photo (cf Fig. 8 for historical photo). Note that the land that was well underway to being cleared just over a century earlier is now cleared, green, and bears little trace of its earlier forest origins. The dead, ring-barked trees in the old photo of Hoschke's farm (Fig. 7) are gone, but the new house with the red roof is in a similar location to the wooden house of a century earlier. Also noticeable is that the sharp line of farm and forest, evident during the initial clearing phase, is now even sharper. The only regrowth is on the riparian strip.

Proc. Linn. Soc. N.S.W., 138, 2016

43

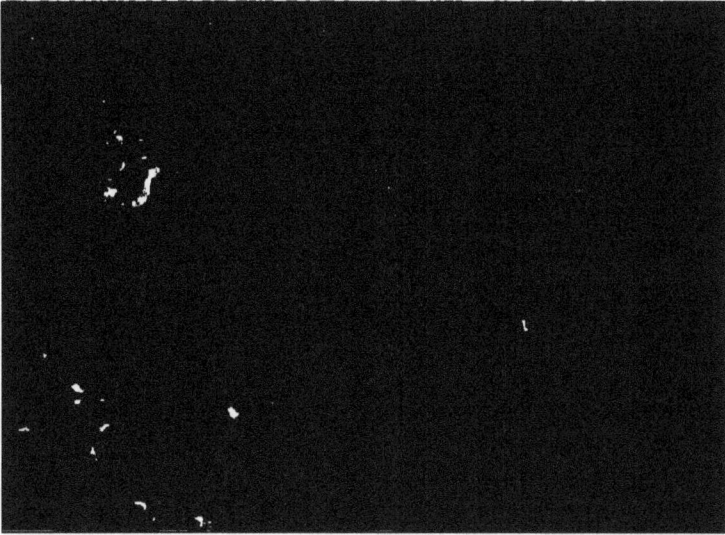

Fig. 2. This 2009 high-resolution aerial photo, in the ADS40 series, shows the present-day location of Cochrane's farm. The most noticeable features of this photograph are the regrowth along the banks of the Orara River, the disappearance of the ring-barked trees, and the stumps. From a Koala ecologist's viewpoint, this is a fragmented and much transformed landscape that would have been prime Koala habitat.

Fig. 3. Bridge across the Orara River, on original site of Hoschke's farm on the other side of the river. Doug Hoschke, grandson of the original farm owner, took Dan Lunney to this site, as he knew both the old photo and the site of what was his grandfather's farm. The farm is no longer in the family. Note the regrowth along the river bank, the cleared land in the background, and the forest on the hills. Photo by Dan Lunney (3 January 2014).

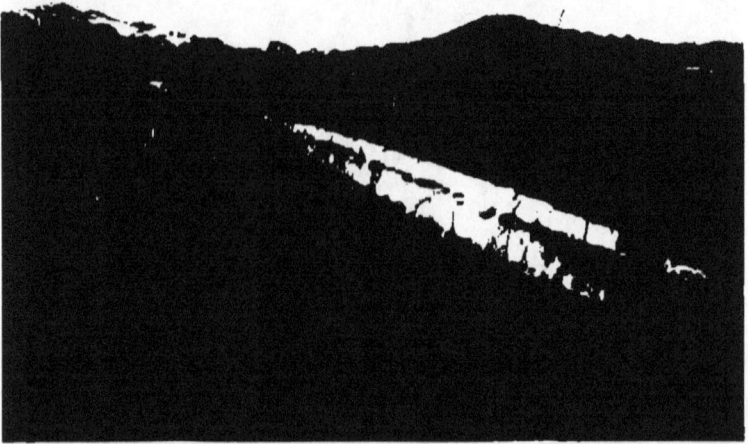

Fig. 4. Doug Hoschke standing on the original site of McLeod's farm. Comparing this photograph to Fig. 8 (historical photo of site) allows us to discern that the modern appearance of the landscape took shape at first settlement, and it has remained very similar today. Photo by Dan Lunney (3 January 2014).

Fig. 5. Contemporary view of the original site of Cochrane's farm. There is now a thin strip of trees growing alongside the riverbank; the land remains cleared in the area adjacent to this strip of trees. In view of identifying Koala habitat, the scene is very similar to that of a century earlier. Photo by Dan Lunney (3 January 2014).

Fig. 6. Contemporary view of the Upper Orara Road, which runs close the Orara River, between the original sites of Hoschke's and Cochrane's farms. Doug Hoschke pointed out to Dan Lunney that the forested slope in the background is regrowth forest that has developed in Doug's lifetime, i.e. since the late 1930s, and Koalas now occasionally occupy this site. However, Doug Hoschke also pointed out that the Koalas cross the road, and are killed on the road. Photo by Dan Lunney (3 January 2014).

Fig. 7. Contemporary view of a bridge across the Orara River linking the original sites of Hoschke's farm (foreground) and McLeod's farm (background). The primary difference between this photograph and those taken over a century earlier is the growth of a thin strip of trees along the river edge. Otherwise, the farmland that was cleared within decades of first settlement has remained cleared farmland.

APPENDIX 2

Fig. 1. Coffs Harbour High School, next to the Coffs Harbour Jetty Post Office, on Harbour Drive, Coffs Harbour, located within the circle #3, in Fig. 10. The High School was the site of Nicholl's timber mill in Fig. 6. Photo by Dan Lunney (3 January 2014).

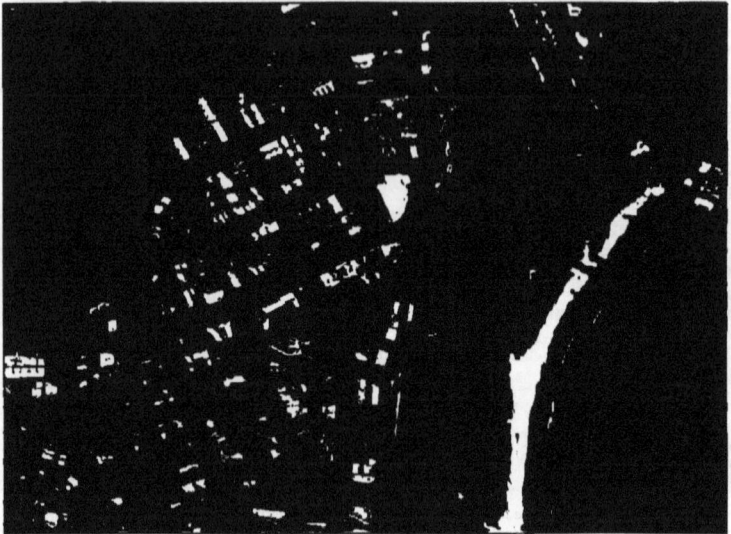

Fig. 2. This modern high-resolution aerial photo, in the ADS40 series, shows the present-day location of Nicholl's timber mill, Fig. 6. Near the centre is Coffs Harbour High School, marked by a cluster of red buildings.

Proc. Linn. Soc. N.S.W., 138, 2016

47

Fig. 1. This photo of modern-day Brelsford Park, Coffs Harbour, shows features identifiable in photos taken over a century earlier. The shape of land, the size and colour of the trees, and the forested hills in the background all help to interpret the use of land at first settlement. Photo by Dan Lunney (2 January 2014).

Fig. 2. The cluster of trees on City Hill, visible to the east of Brelsford Park, Coffs Harbour, is remnant Koala habitat, and modern records of Koalas at this location exist. This photo, combined with earlier photos, modern Koala surveys, and early records, confirms that Koalas were found and still are to be found in Coffs Harbour. Photo by Dan Lunney (2 January 2014).

48

Proc. Linn. Soc. N.S.W., 138, 2016

Julian Tenison Woods: Natural Historian

Presidential Address delivered at the 141st Annual General Meeting of the Linnean Society of New South Wales, March 23rd 2016 by Emeritus Professor Robert J. King.

Published on 12 July 2016 at http://escholarship.library.usyd.edu.au/journals/index.php/LIN

King, R.J. (2016). Julian Tenison Woods: natural historian. *Proceedings of the Linnean Society of New South Wales*. 138, 49-56.

'The time has again come around when the duty devolves upon me to deliver the Annual Address to the members of the Society'. With these words the Rev. Julian Edmund Tenison Woods commenced his 1880 Address at the conclusion of his second term as President of the Linnean Society of New South Wales.

Tonight I wish to provide a brief introduction to the scientific works of Father Tenison Woods and to highlight his contributions to the study of Australian natural history. Tenison Woods has in recent years come to public attention because of his close association with Saint Mary of the Cross, Mary McKillop, with whom he was associated in the early years of the Sisters of Saint Joseph. For those interested in an account of the life of Tenison Woods, his work as a priest and his problematic relationships with the Catholic Church hierarchy, I refer you to the second edition of 'Julian Tenison Woods: Father Founder' by Margaret M. Press (Press, 1994). In the present paper I do not wish to address those aspects, which are clearly out of my area of expertise. What I want to do is provide an overview of the very major contribution made by Tenison Woods to early studies of natural sciences in Australia. I cannot hope to cover in detail the numerous papers he published in the period from the 1850s to his death, at the age of 56, in 1899. A detailed bibliography, including a list of published scientific writings, can be found in an Appendix to Margaret Press' biography.

Julian Woods was born in London into a large family of 11 children, a family that encouraged learning and especially the study of nature. His education appears to have been quite haphazard and he attributed much to his father, especially his interest in history. His father, James, was a member of the Society of Antiquaries. Woods developed a lifelong habit of reading, of self-education and an interest in a broad range of disciplines. He spent much of his late adolescence and early adulthood seeking his vocation,

and this included a brief period in France, where he had hopes of improving his health. This concern for his health played a role in his decision to migrate to Australia. In 1854 he set out for Tasmania arriving early in 1855, but disappointed in what was offering there he left to visit relatives in Melbourne, before moving to Adelaide towards the end of 1855. He was eventually ordained a priest in Adelaide in 1857, and a few months later began his work in the parish of Penola, in southeast South Australia. In 1866 he and Mary MacKillop founded the Sisters of St Joseph, dedicated to the education of the Catholic poor and others with social needs. Later that year he was appointed Director General of Catholic Education in Adelaide, a position he held for some four years. After he was eased out, or perhaps actively 'moved on', he worked as a missionary priest in New South Wales, Tasmania and Queensland. He continued to have difficulties with his superiors but remained in this role until 1883 with little apparent support, and in some cases active opposition, from the Church hierarchy.

Before discussing the scope and significance of the research undertaken and published by Tenison Woods it is appropriate to consider the special attributes that he was able to bring to his scientific life.

- His background as a keen observer of natural history in both England and briefly in France, was sufficient for him to perceive similarities and differences in the fossils of different localities, especially in the Tertiary fossil faunas. He did not interpret the Australian situation in isolation; he saw the bigger picture. In one of his earliest papers on fossils in the limestone at Mount Gambier he alluded to similarities with fossils in the chalks of the Upper Crag in Suffolk, England.
- He grew up in a household where his father worked as the parliamentary reporter for *The*

Times and at various stages in both England and Australia Tenison Woods found employment with the press. This presumably gave him confidence and connections when it came to publishing some of his material, especially in the earlier period when there were few relevant journals being published in Australia.

• He was courageous in publishing in areas of Australian natural history where there was little or no background, but he acknowledged these shortcomings.

• He maintained contact with key researchers, initially in Europe. Especially key here was encouragement from Charles (later Sir Charles) Lyell, the father of geology. His interactions with scientists in Australia were, I believe, key to his influence in the developing colony. His extended period as a missionary priest in eastern Australia including Tasmania and later the Northern Territory allowed him to make personal contact to mutual advantage and this is well exemplified in his relationship with F. M. Bailey, with whom he published on botany in Queensland.

• He was hard working with an enviable scientific output despite his commitment to church matters. This aspect was alluded to in his own writing, but also that of colleagues.

• He had a special capacity to read and synthesise material, and to collate it in readable form. This is particularly evident in his book on the 'Fish and Fisheries of New South Wales', to which I will refer later.

Tenison Woods had a lifelong commitment to public education. This was demonstrated from his time in Penola where he saw the education of the children of the poor as essential, and throughout his later period as a public intellectual and scientist. It is difficult to separate his zeal for public education and his enthusiasm for science. He regretted the low status of fundamental science in society. In 1880, in words which are of particular relevance today, he wrote 'Scarcely a meeting or a public discussion is there in which some daunting allusion is not made to the progress of knowledge and our intellectual achievements. This as far as it goes, is a sign of some sort of appreciation in which the labours of a few are held' but 'Science and scientific study are not popular. Scientific results, when they benefit mankind, are appreciated and admired, men of science, when their reputation is established hold a high and honourable position; but the labour by which all this is acquired has very few votaries indeed'. Tenison Woods did as much as anyone to address this problem of the image

of science and scientists through public lectures (in 1865 for instance two lectures on the geology of Portland Victoria: and on leaving Penola a summary of his natural history observations was made in a lecture entitled 'Ten Years in the Bush'). He also published a large number of letters and commentaries in the press.

In assessing Tenison Woods' contribution to various disciplines, the scientific environment in which he found himself needs to be taken into account. In broad terms, the 19th Century studies of natural history in Australia fall into three categories. In the first period specimens were collected on exploration voyages and returned to Europe for their scientific investigation, in the second collections were made by Australians but generally described overseas, and in the third period they became the subject of Australian studies. Tenison Woods' earliest scientific work is in the second period; for instance he made a large collection of fossils and sent them to Britain, to Sir Charles Lyell who much encouraged him with his geological studies. Tenison Woods went on to become one of the key players in the third phase, collecting and then himself describing new taxa based on those collections. Because he collected material in the field, rather than working on necessarily limited collections, Tenison Woods developed a keen understanding of plasticity within the species and speculated on variation attributable to both biogeographical aspects and local changes of environment. This meant that he did not describe as new taxa every variant he came across. This is well demonstrated in his work on littorinids, or periwinkles.

In his Presidential address to the Society in 1880 (Tenison Woods, 1880) he referred with pride to the 'labours of scientific men in the colonies', many of whom he knew personally, and he addressed the difficulties they encountered with access to the relevant literature. He commented that when he made his earliest studies on Tertiary fossils in South Australia the written works of the key European palaeontologists were (not unexpectedly) 'not accessible in the Australian bush'. He drew attention to the problem of access to scientific papers generally, noting that much was scattered through the scientific journals of Europe or attached as appendices to works on the colonies. 'How few for instance, have seen Dr Lindley's papers on the flora of Western Australia or Stutchbury's remarks on the Natural History of Port Jackson. Would any library in Australia be likely to contain the Proceedings of the History Society of Metz, with Arthur Morellet's descriptions, or how difficult it would be to obtain Menke's Latin pamphlet on the Mollusca of New Holland, published

50

Proc. Linn. Soc. N.S.W., 138, 2016

in Hanover. A valuable pamphlet of Menge's on the Mineralogy of South Australia is as difficult to meet with as an Elzevir Sallust' (published by the House of Elzevirin 1634).

This situation underscores Tenison Woods' commitment to, and recognition of, the important role that societies such as the Linnean Society of New South Wales played in publication of science of local interest and importantly in the 1800s in making the results readily available in Australia. He noted in his 1881 President's Address that during the past year the Society had 'issued a volume which will bear comparison with any scientific serial for the extent and importance of the matter contained'. He also congratulated the Society on seeing the necessity of having some rooms and a library of its own. Tenison Woods was also reputed to have a significant personal library.

Julian Woods' first major scientific work was 'Geological Observations in South Australia: principally in the district south-east of Adelaide' published in 1862. He was conscious of the need to establish his reputation and at some stage, certainly by 1866, had begun using his third given name (Tenison, the maiden name of his mother) to distinguish himself from two other natural scientists surnamed Woods. Thereafter all of his scientific papers used the appellation Tenison Woods, hyphen or not.

Tenison Woods published on a wide range of topics, a breadth that for a researcher in the 21st Century is unimaginable. He was an astute observer of natural history and made significant original contributions in geology, palaeontology, botany and zoology. His interests were perfectly aligned with those of the Linnean Society of NSW. His breadth of scholarship, however, encompassed far more than the sciences and in addition to Church matters he published on history, bibliography and more. His early reputation was established in large part on essentially non-scientific works, with the publication of his 'History of the Discovery and Exploration of Australia' in two volumes in 1865 (even if the Geological Observations in South Australia had been published three years earlier), and his 'Australian Bibliography', a serialised survey published in the Australian Monthly Magazine from 1866 – 1867. Some of his most important scientific publications affirm his great capacity to synthesise material and to present it in a cogent manner, rather than the creation of new knowledge. This is most evident in his major book 'Fish and Fisheries of New South Wales' published in Sydney in 1883. This major work is of such significance that it has been republished as a 'forgotten book'. The book was commissioned by

the Colonial Government as a complete handbook of the fish and fisheries and was designed to promote development of this resource. It was to accompany the New South Wales Exhibition at the Fisheries Exhibition in London in 1883. This commission came at a propitious moment when his major source of income through missions was no longer available. In this we can perhaps see the hand of William Macleay (perhaps the nearest thing the Linnean Society of New South Wales has to a 'father founder') whose investigations into ichthyology were, to quote Tenison Woods, 'given most distinguished votaries'. In the fisheries book Tenison Woods' skills in writing for a general audience are beautifully exemplified. Recognition of the value of this publication and other works came from the unusual source when King William III of the Netherlands awarded him a gold medal for the best publication of the Exhibition, when the treatise had been translated into Dutch for the Amsterdam Exhibition in 1884.

What I've said up to now has been of a general nature. I now want to turn my attention to Tenison Woods' research papers. While initially these were largely focussed on geology he later expanded his published research to physical geography and natural history more broadly. Geology remained his favourite interest and embraced palaeozoology, with a particular interest developed in marine Mollusca and Bryozoans. By the late 1870s he had also published extensively on corals, echinoderms, and land snails. His interests broadened considerably thereafter, such that in 1879 he was publishing on the vascular flora (distribution and biogeography), fungi and lichens. By the early 1880s his developing interest in fossil flora and coal deposits came to the fore. This aspect, and an interest in mineralogy generally, is reflected particularly in the papers in the latter part of his scientific career when he travelled and made observations in northern Australia and in southeast Asia, especially Malaya.

In September 1989, the Centenary of Tenison Woods' death, the Earth Sciences History Group of the Geological Society of Australia Inc. in Sydney organised a symposium on the scientific work Tenison Woods. The symposium was strongly supported by the Sisters of Saint Joseph, who gave generous financial assistance. Papers from that meeting were published, by the Royal Society of New South Wales, in 1991. The comments that follow are largely based on those papers and directed to Tenison Woods' contributions to geological studies, where he had his greatest impact. Before I do that however, and as a botanist, I should make a comment on his later botanical publications. Peter Martin, in the papers resulting from the 1989 symposium, wrote that Tenison Woods was 'a highly

JULIAN TENISON WOODS: NATURAL HISTORIAN

competent botanist. His published papers on modern and fossil botany would, by themselves, have been sufficient to establish him as a significant figure in the annals of Australian science'. In my view this assessment applies to his work on fossil floras only. His botanical work on living species was undertaken largely in Queensland, often in conjunction with F. M. Bailey and was not in any sense ground-breaking. In those papers Tenison Woods did little more than extend the knowledge of the distribution and habitat of some species.

I now turn my attention to the scope and significance of geological research undertaken by Tenison Woods and in doing so wish to acknowledge my debt to Dr Ian Percival, for his commentary and advice.

In 1889, a few months before the death of Tenison Woods, C. S. Wilkinson (NSW Government Geologist and President of the Royal Society of NSW) wrote in his Address for the Clarke Medal - named after another famous Reverend and geologist - that geology was Tenison Woods' 'favourite branch of Science'.

Tenison Woods' geological studies are readily divisible into geographical regions, chiefly South Australia and Victoria, Queensland, New Guinea and the Pacific islands, and Southeast Asia (principally Malaya and the Dutch East Indies), reflecting the areas where he spent sufficient time to develop an interest in the local rocks, fossils, coal and mineral deposits, and landforms. One can't help feeling that Tenison Woods' unusual relationship with power structures in the Church, often moving around the country, played out to the benefit of science and exploration. The regional interests also shaped his studies into fossils of particular ages, chiefly Tertiary of the southern Australian mainland and Tasmania but including the preceding Mesozoic Era in Queensland. He sometimes delved into specific geological problems such as the origin of the Hawkesbury Sandstone in the Sydney Basin, and published observations on the geology and mineral potential of the Northern Territory (at the time, part of the colony of South Australia). His research therefore had a considerable geographic spread, and included pioneering studies of aspects of regional geology as diverse as fossils, caves, volcanoes, coal and ore deposits, and hydrogeology. A very 21st Century way of assessing whether he left an enduring legacy in any of these fields would be to check citations of his work in the most recently published compilations of the geology of the various states and territories of eastern and central Australia – and on this criterion it could be said that his impact has been largely forgotten or superseded, except in the study of Tertiary palaeontology. Perhaps the same

assessment would be made of Charles Darwin or Galileo! As Archbold noted at the 1989 symposium: 'Many of his taxa have survived the subsequent century of study'. Of some 20 species of Bivalvia and 120 species of Gasteropoda named by Woods from Tertiary strata only 3 and 7 respectively have not survived.

Tenison Woods' earliest published scientific studies were made in South Australia and Victoria. Halfway through the decade (1857-1867) that Tenison Woods spent in South Australia, he published his book on 'Geological Observations in South Australia: principally in the district south-east of Adelaide' (1862). This contained descriptions of the volcanic landforms and crater lakes of Mount Gambier, the limestone caves of that area and those at Naracoorte, and notes on the abundant fossils of the Tertiary (Miocene) age of this region, including those from the Murray River cliffs. From these strata in the vicinity of Mannum, Tenison Woods described in 1862, the first fossil echinoid from Australia, referred by him to *Spatangus* and now known as *Lovenia forbesii*. Many of these fossils are identical to those he later described from Victoria, especially the fauna from Muddy Creek near Hamilton, Batesford Quarry near Geelong, and Fossil Beach on the Mornington Peninsula. He recognised the faunal similarities and correlation of these widely separated localities.

In Tasmania, from 1874 until early 1877, Tenison Woods studied the rich Tertiary faunas at the appropriately named Fossil Bluff, near Wynyard in northwest Tasmania. He described at least nine taxa of molluscs of early Miocene age from this locality.

Tenison Woods returned to Sydney in 1877. In a paper at the 1989 symposium Kevin McDonnell assessed Tenison Wood's study on the Hawkesbury Sandstone as one of his major contributions, noting that it provided clear testimony to his considerable stature as a scientist and pioneer Australian geologist. McDonnell wrote of this work: 'His interpretation of the Hawkesbury Sandstone as a wind-blown formation is supported by his observations of its geometry, lithology, sedimentary structures and fossil content; by comparison with aeolian and other formations in Australia and in various other parts of the world, either through the literature or by personal observation; by experiments he conducted with wind-blown sand, and by personal observation of aeolian processes in the field. Although his interpretation of the origin of the Hawkesbury Sandstone as a whole is not accepted today (he did not have available to him the detailed knowledge we now have of the processes and products of fluvial and other environments) his method was sound and his competence undoubted'. I've quoted

52

Proc. Linn. Soc. N.S.W., 138, 2016

this passage in full because it beautifully illustrates some of the generalisations I made earlier: his breadth of approach (geometry, lithology, sedimentology and palaeontology), his experiences allowing comparison with other formations in Australia and world-wide, his attention to the broader literature, and his perceptive field observations.

Following his return to New South Wales Tenison Woods made intermittent visits to Queensland between 1878 and 1883 and these were particularly important in spurring his interest in botany, both extant flora and palaeobotany. Previously he had confined his fossil studies primarily to invertebrate palaeontology. While investigating the coal resources of the colony on a government-funded project, he described elements of the associated *Dicroidium* flora of Triassic age from Ipswich. He also described Cretaceous ammonites and a belemnite from the Walsh River region of northwest Queensland.

At the end of 1882 Tenison Woods had reached some sort of crisis in his relationship with the Catholic bishops and he was required to cease giving missions or officiate in a number of dioceses. At this time he had received a tempting invitation from Sir Frederick Weld, an old Tasmanian friend who was now Governor of the Straits Settlements and living in Government House in Singapore. In this last stage of his geological career, Tenison Woods concentrated more on documenting coal and mineral deposits, continuing the trend commenced in Queensland where he had investigated coal and tin mining areas. Much of this work, both in Queensland and in southeast Asia, was undertaken on commission from government officials. He made observations on the Malay Peninsula, Malacca, Java, Borneo, China, Siam and Japan, and published some of the research in our Proceedings, and general observations in the press.

Following his return to Australia, in the brief period when his health permitted, he further explored the mineral districts of the Northern Territory. In 1885, Tenison Woods speculated that Arnhem Land would become one of the greatest mining centres in Australia. How prescient, even if he didn't know why.

Tenison Woods returned to Australia in 1886, but it was a lengthy journey, and landing in Port Darwin he had the opportunity to visit Victoria River and then undertake some geological survey work on behalf of the South Australian Government. By this time his health, always an issue, was worsening. As long as he was able he wrote up his recent exploration notes.

In his foreword to Margaret Press's biography of Julian Tenison Woods, Paul Gardiner wrote 'There will always remain a question-mark over the spiritual insights which led the gifted founder (of the Institute of Saint Joseph) to act as he did. At times the evidence points to definite weaknesses in his mental processes. These found expression in astonishing language and led to some bizarre courses of action. Woods puzzled his contemporaries'. If Tenison Woods puzzled his contemporaries in the Catholic Church he did not puzzle his scientific friends and colleagues. At the time of his death Professor Archibald Liversidge aware of Tenison Woods' scientific repute, praised his 'great simplicity, courtesy and kindness of manner'; and J. C. Cox, Wood's successor as President of our Society, testified to his 'exuberant industry ... [and] extraordinary variety of attainments'. His memorial in the Waverley Cemetery in Sydney is a fitting tribute (Fig. 1). Erected with public funding, the greatest contributions are reputed to have come from his scientific colleagues.

Tenison Woods received well-deserved recognition in his lifetime. In the year before he died he was awarded the 1888 Clarke Medal of the Royal Society of New South Wales for his natural history

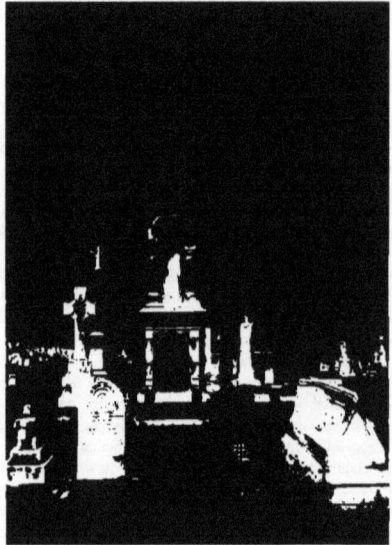

Fig. 1. Memorial to Tenison Woods, Waverley Cemetery, Sydney.

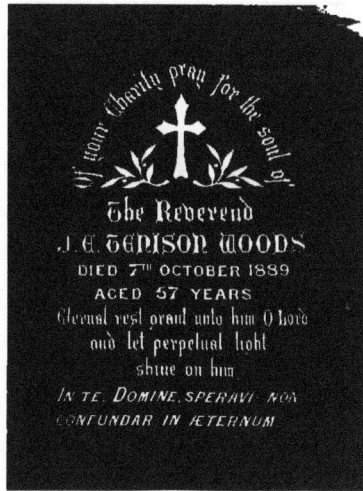

Fig. 1 cont'd. Plaques on the sides of the memorial to Tenison Woods, Waverley Cemetery, Sydney

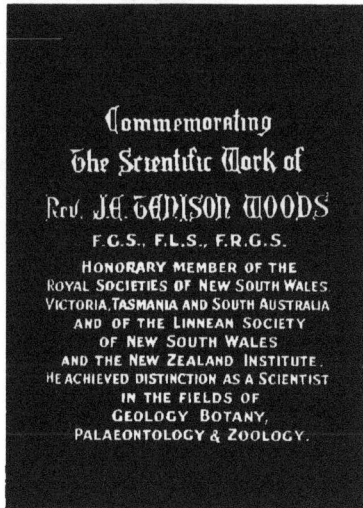

works generally, but particularly for his geological studies. The Clarke medal, is awarded 'for meritorious contributions to Geology, Mineralogy and Natural History of Australasia, to be open to men of science, whether resident in Australasia or elsewhere'. He was the 11th recipient of the Clarke Medal, joining an illustrious group including among others George Bentham, Thomas Huxley, Baron Ferdinand von Mueller, and Sir Joseph Dalton Hooker. He was the only member of the distinguished group not to hold or have held a government scientific post.

Tenison Woods received the Passionist habit on his deathbed, and if the pun can be excused he was also passionate to the end about his science and proud of his membership of scientific societies. He was able in 1887 to list honorary membership of the Royal Society of New South Wales, the Royal Society of Tasmania, the Royal Society of South Australia, the Straits Branch of the Royal Asiatic Society, the Royal Geographical Society of Queensland and New South Wales, the New Zealand Institute, the Microscopical Society Victoria, the Field Naturalists Club of Victoria, and he was a corresponding member of the Royal Society of Queensland and of the Royal Society Victoria. He was also a fellow of the Geological Society of London. The Linnean Society of New South Wales however claims his greatest allegiance: he was admitted to membership of the Society in 1876

54

Proc. Linn. Soc. N.S.W., 138, 2016

Fig. 2. Tenison Woods, President of the Linnean Society of New South Wales, 1879-1881.

and was President of the Society in 1879 and 1880 (Fig. 2), and thereafter Vice-President until his death. He published some 70 papers in our Proceedings.

It would be inappropriate to talk about any 19th Century natural historian without some reference, however brief, to Charles Darwin. Darwin and Tenison Woods shared the distinction of being honorary members of the Royal Society of New South Wales. Tenison Woods had clearly read Darwin's 1842 monograph on 'The Structure and Distribution of Coral Reefs'. We know this because he recorded that he was not in complete agreement with all of the conclusions therein. In his 1880 Presidential Address he commented at length on Darwin's book 'Effects of Cross- and Self-Fertilisation in the Vegetable Kingdom' published in 1876, noting especially the impetus to further research the book had inspired. Given the comment in the Australian Dictionary of Biography that Tenison Woods possessed 'profound, and romantic religious convictions based on a childlike piety' one might have wondered about his response to Darwin's views on evolution. There is, however, no

need to speculate. In various places in his 1880 Presidential Address Tenison Woods commended Darwin as 'ingenious', 'conscientious', 'illustrious' and noted the 'perfection of his methods of enquiry'. He concluded 'I can well believe that there is much truth in evolution. If tomorrow the evidence of its occurrence were established on indubitable grounds, it would be one more beautiful illustration of the plan of nature'. It is perhaps worth noting that his views were entirely consistent with those that the Roman Catholic Church finally propagated in the 1950 Encyclical which confirmed no intrinsic conflict between Christianity and the theory of evolution, in other words theistic evolution. That view was almost a century in its formulation.

How will Tenison Woods be remembered? In addition to Tenison Woods' role as it relates to Mary MacKillop (now formally known as Saint Mary of the Cross) he will be remembered always for his key role in Roman Catholic school education, especially directed to the poor and needy. Two schools in South Australia are named in his honour: Tenison Woods Catholic School, an R-7 primary school, in western Adelaide, and Tenison Woods College, an Early Learning to Year 12 Catholic Co-Educational College located in Mount Gambier. His contribution to geography is recognised in the naming of Mount Tenison Woods, the highest point in the D'Aguilar Range near Brisbane.

In science it is for his pioneering role in a number of branches of study and for his advocacy of science that he should be best remembered. By spreading his studies and publications in geology, and natural history more generally over so many subjects, rather than specialising in just one or two fields, Tenison Woods never really became the recognised authority in any area, other than Tertiary palaeontology in which he excelled. Much of his geological work has been superseded by subsequent observations and discoveries and is now largely overlooked. This is particularly true of his paper on the origin of the Hawkesbury Sandstone, widely regarded at the time of its publication as among his best works. However, Tenison Woods has not been entirely forgotten, especially by palaeontologists. He has been commemorated in the names of at least eight fossil taxa and some extant species (even a higher plant, *Leucopogon woodsii*). Ian Percival has provided the following list of fossils named in his honour, including the genera *Jetwoodsia* (a gastropod), *Tenisonina* (a foram) and the species *woodsi, woodsii* and *tenisoni*. His unusual double barrelled name expanded the possibilities.

Austrotriton woodsii (Tenison Woods, 1879) Batesford Quarry, Geelong, Vic
Jetwoodsia apheles (Tenison Woods, 1879) Muddy Creek, Hamilton, Vic
Belaphas woodsii (Tate, 1888)
Lovenia woodsi (Etheridge, 1875) Murray River Cliffs, Sunlands, SA, Loxton Sands Formation
Terebra tenisoni (Finlay, 1927)
Mopsea tenisoni (Chapman, 1913)
Jetwoodsia nullarborica (Chapman & Crespin, 1934)
Tenisonina tasmaniae Quilty, 1980 (Early Miocene foraminiferid from Fossil Bluff, Tas.)

As I come to the conclusion of this address I wish to return to Tenison Woods himself and the hopes he had for the future of our Society. He referred specifically to the helping hand that we can extend to 'students of science, especially beginners'. This aspect has now developed as one of this Society's main objectives, with grants available through several bequests and donations. I am delighted to acknowledge the very generous donation from the Sisters of Saint Joseph as a practical way marking the Sesquicentenary of the Foundation of the Congregation by Father Julian and Mary McKillop on March 19, 1866. Through their generous donation the Linnean Society of New South Wales will fund research in any one of the fields in which Tenison Woods made his contributions to the natural sciences.

My last words are fittingly from the 1881 Presidential Address by the Rev J. E. Tenison Woods as he stood down from the Presidency and took up a position as Vice President. 'I must again congratulate my fellow workers in this Society on their industry and zeal. They have laboured so indefatigably that I can look back to the period of my Presidency as one which has largely added to the reputation for usefulness and efficiency which the Linnean Society has gained'.

REFERENCES

Press, M. M. (1994) 'Julian Tenison Woods: Father Founder', Second Edition, Collins Dove, Melbourne. 270pp.
Tenison Woods, J. E. (1880) President's Address, Annual Meeting 28th January 1880, Proc. Linn. Soc. NSW, 4: 471-491.
Tenison Woods, J. E. (1881) President's Address, Annual Meeting 27th January 1881, Proc. Linn. Soc. NSW, 5: 638-652.

New Information about the Holotype, in the Macleay Museum, of the Allied Rock-wallaby *Petrogale assimilis* Ramsay, 1877 (Marsupialia, Macropodidae)

Graham R. Fulton

School of Veterinary and Life Sciences, Murdoch University, South Street, Murdoch WA 6150, Australia
Centre for Biodiversity and Conservation Science, The University of Queensland, Brisbane Qld 4072, Australia
(grahamf2001@yahoo.com.au)

Published on 12 July 2016 at http://escholarship.library.usyd.edu.au/journals/index.php/LIN

Fulton, G.R. (2016). New information about the holotype, in the Macleay Museum, of the Allied Rock-Wallaby *Petrogale assimilis* Ramsay, 1877 (Marsupialia, Macropodidae). *Proceedings of the Linnean Society of New South Wales* **138**: 57-58.

The location of the missing holotype of the Allied Rock-wallaby *Petrogale assimilis* is given as the Macleay Museum. Background information about its collection during the *Chevert* Expedition of 1875, obtained from Sir William Macleay's personal journal, sheds further light on the history of this important specimen.

Manuscript received 12 April, accepted for publication 25 June 2016.

Keywords: *Chevert* Expedition, Macleay Museum, Nyawaygi people, type specimens.

DISCUSSION

Type specimens serve as the physical reference point for a particular taxon. Most holotypes are single specimens upon which the description and name of a species is based (ICZN 1999). The holotype for the Allied Rock-wallaby *Petrogale assimilis* was established by Edward Pierson Ramsay in March 1877. In reference to this specimen he wrote, "I believe, the only specimen obtained; sex, female" (Ramsay 1877a. p. 360). Ramsay was referring to the *Chevert* Expedition headed by Sir William Macleay. Macleay's personal journal shows that the specimen was collected on June 2, 1875 at Palm Island (18°43'55.9"S 146°36'22.5"E), Queensland. Palm Island falls within the provenance of the Nyawaygi people (Horton 1996). A little later, in July 1877, Ramsay published another note on this species. This time Ramsay stated two specimens were collected "an adult and a young" and gave only a very brief description of the fur on the younger specimen (Ramsay 1877b. p. 11). The months and year of Ramsay's publications were determined from Joseph James Fletcher's publication of the dates for early issues of the *Proceedings of the Linnean Society of New South Wales* (Fletcher 1896).

The two specimens referred to by Ramsay (1877a. 1877b) were collected during the *Chevert* Expedition in 1875 (Fulton 2012). Macleay's collectors George Masters, Edward Spalding and Dr W. H. James collected the specimens, inland on the island, with the aid of an unnamed Nyawaygi guide (Macleay 1875). The specimens were incorporated into the collections of the Macleay Museum, which were subsequently donated, along with its building, to The University of Sydney. The University placed the collection into storage soon after Macleay's death to make use of the building for other purposes (Fulton 2012). Approximately 80 years later many of the Museum's type specimens were sent to other institutions. The mammals were thought to have been moved to the Australian Museum along with the birds (Stanbury 1969a, 1969b). Fulton (2001, 2012) found some birds were missing and they might still be in the Macleay Museum.

The Australian Faunal Directory, an incomplete online catalogue of taxonomic and biological information on all Australian animal species, registered the following status for the type data on *Petrogale assimilis*: "Holotype whereabouts unknown, Palm Is., N of Townsville, QLD" (ABRS 2009). Upon a search at the Macleay Museum the two specimens were found, M422 (female holotype) and M423 (male). They are currently labelled NHM.422 and NHM.423.

Given this discovery, it is possible that other type specimens whose whereabouts is currently unknown may reside in the collections of the Macleay Museum.

ACKNOWLEDGEMENTS

I thank Jude Philp and Robert Blackburn for their necessary assistance at the Macleay Museum, because I am at the other end of the continent. I thank Stephen Jackson for comment on a draft manuscript. I acknowledge the Nyawaygi people the traditional owners of Palm Island where the Allied Rock-wallaby was collected and I acknowledge their help in originally collecting the specimens.

REFERENCES

ABRS (2009). *Petrogale assimilis*. Australian Faunal Directory. Australian Biological Resources Study, Canberra. http://www.environment.gov.au/biodiversity/abrs/online-resources/fauna/afd/taxa/Petrogale_assimilis (Viewed January 5, 2016).

Fletcher, J. J. (1896). On the dates of publication of the early volumes of the Society's Proceedings. *Proceedings of the Linnean Society of New South Wales*, **10**, 533-536.

Fulton, G. R. (2001). Threatened and extinct bird specimens held in the Macleay Museum, University of Sydney, Australia. *The Bulletin of the British Ornithologists' Club*, **121**, 39-49.

Fulton, G. R. (2012). Alexander, William Sharp, and William John Macleay: Their Ornithology and Museum. Vol. 2: 327-393. In: 'Contributions to the History of Australasian Ornithology Vol. 2.'(Eds. W. E. Davis, Jr., H. F. Recher, W. E. Boles and J. A. Jackson.) pp. 327-393. (Cambridge, Massachusetts: Nuttall Ornithological Club).

Horton, D. R. (1996). Indigenous Language Map. Aboriginal Studies Press, AIATSIS and Auslig/Sinclair, Knight & Merz. http://www.abc.net.au/indigenous/map/default.htm (Viewed January 5, 2016).

International Commission on Zoological Nomenclature (1999). 'International Code of Zoological Nomenclature. Fourth Edition.' Online. (Viewed January 5, 2016).

Macleay, W.J. (1875). Personal journal transcribed by D.S. Horning in 1995. Transcript held in the Macleay Museum, University of Sydney.

Ramsay, E. P. (1877a). Description of a supposed new species of Rock Wallaby from the Palm Islands; on the north-east coast of Australia, proposed to be called *Petrogale assimilis*. *Proceedings of the Linnean Society of New South Wales*, **1**, 359-361.

Ramsay, E. P. (1877b). Zoology of the "Chevert". Mammals. Part 1. *Proceedings of the Linnean Society of New South Wales* **2**, 7-19.

Stanbury, P. J. (1969a). Type specimens in the Macleay Museum, University of Sydney. III. Birds. *Proceedings of the Linnean Society of New South Wales*, **93**, 457-461.

Stanbury, P. J. (1969b). Type specimens in the Macleay Museum, University of Sydney. IV. Mammals. *Proceedings of the Linnean Society of New South Wales*, **93**, 462-463.

Bramble Cay Melomys *Melomys rubicola* Thomas 1924: Specimens in the Macleay Museum

GRAHAM R. FULTON

School of Veterinary and Life Sciences, Murdoch University, South Street, Murdoch WA 6150, Australia
Centre for Biodiversity and Conservation Science, The University of Queensland, Brisbane Qld 4072, Australia
(grahamf2001@yahoo.com.au)

Published on 12 July 2016 at http://escholarship.library.usyd.edu.au/journals/index.php/LIN

Fulton, G.R. (2016), Bramble Cay Melomys *Melomys rubicola* Thomas 1924: specimens in the Macleay Museum. *Proceedings of the Linnean Society of New South Wales*, **138**, 59-60.

Four specimens and a lower mandible in spirit of the Bramble Cay Melomys *Melomys rubicola* were recently found in the Macleay Museum. These specimens were collected during the *Chevert* Expedition in 1875 and were not published as part of the mammals obtained. The species is now considered extinct. An old newspaper article written by the ship's captain, Charles Edwards, provided the clue that this extinct species was extant in the Macleay Museum. The DNA of surviving specimens may yet provide the answer to the origin of the endemic Bramble Cay Melomys.

Manuscript received 20 February 2016, accepted for publication 25 June 2016.

Keywords: Bramble Cay, Captain Charles Edwards, *Chevert* Expedition, climate change, Erubam Le people, Macleay Museum, *Melomys rubicola*, Torres Strait.

DISCUSSION

Bramble Cay is a small vegetated sand cay of about 5 ha surrounded by a coral reef and located in north Torres Strait (9°08'31.1"S 143°52'29.9"E), approximately 50km from the mouth of Papua New Guinea's Fly River. Bramble Cay falls within the provenance of the Erubam Le people. The High Court of Australia, in 2004, granted the Erubam Le native title rights over Bramble Cay (Latch 2008). The Bramble Cay Melomys *Melomys rubicola* was endemic to the Cay and to Australia (Limpus et al. 1983). Its population on the Cay was given at "several hundred individuals" in 1978 (Limpus et al. 1983) and subsequently with an estimated population of 93 in July 1998 (Dennis and Storch 1998). Further declines were recorded leading to an extensive but unsuccessful search in 2014, which found that oceanic inundation associated with human-induced climate change was the root cause of its extirpation from the Cay (Woinarski et al. 2015a, 2015b; Kim and Pressey 2015; Gynther et al. 2016). The Bramble Cay Melomys is thus probably the first mammalian extinction recorded due to anthropogenic climate change (Gynther et al. 2016).

The holotype of *Melomys rubicola* (BMNH 46.8.26.7 ♂ skin & skull) was collected by John MacGillivray in 1845 during the voyage of the H.M.S. *Fly*. Other specimens were collected on the same voyage by Joseph Beete Jukes. The species itself, however, was not formally described until nearly 80 years later in 1924 (Thomas 1924). In the meantime, William Macleay's *Chevert* Expedition had collected four more specimens on the Cay. Alas, these were not reported in the scientific literature along with other mammals collected (e.g., Ramsay 1877a, 1877b). Had they been described at that time they would represent the type specimens of this species. William Macleay did not record collecting this species in his private journal, although he recorded the *Chevert* stopping there and collecting generally on August 13, 1875. The Captain, who was not one of the collectors, supplied a brief narrative of the voyage to a Sydney newspaper later that year. He wrote, "On the 13th, at 7 20, we sailed for New Guinea, touching at Bramble Bay en route. We anchored at 11 in 22 fathoms fair holding ground, and good shelter, with bay bearing S.E. by E. two cable lengths distant. Here we got great numbers of birds and amongst other things, large centipedes, and a rat peculiar to the island"

(Edwards 1875). In fact, based on the Macleay Museum's current collection data they collected four individuals. The four specimens in ethanol: M738 unsexed, one female M739, two males M740 and M471 and a lower mandible labelled M741.

The origin of the Bramble Cay Melomys, in terms of the source population, remains unknown even after its extinction. There are currently two competing theories regarding its origin. One theory suggests that due to the close proximity of the Fly River the Bramble Cay Melomys may have travelled from Papua New Guinea to Bramble Cay on driftwood (Smith 1994). Alternatively, it may be a relict persisting from an earlier the time when Australia was joined to Papua New Guinea by a land bridge (Dennis and Storch 1998). Whatever its origin the DNA preserved in surviving museum specimens may help establish its closest relatives and hence the possible origins of this elegant but little understood species.

ACKNOWLEDGEMENTS

I thank Jude Philp for the data on the specimens, Chris Dickman and Bob Pressey for reading a draft manuscript and Stephen Jackson for helping source the typification. I acknowledge the Erubam Le people the traditional owners of Bramble Cay where the Bramble Cay Melomys were collected.

REFERENCES

Dennis, A. and Storch, D. (1998). Conservation and taxonomic status of the Bramble Cay melomys *Melomys rubicola*. 'Unpublished Report to Environment Australia Endangered Species Program Project No. 598.' (Queensland Department of Environment).

Edwards, C. (1875). Narrative of the *Chevert's* Voyage to New Guinea. *Evening News* 2633 p 2. Thursday, December 9, 1875.

Gynther, I., Waller, N. and Leung, L.K.-P. (2016). Confirmation of the extinction of the Bramble Cay melomys *Melomys rubicola* on Bramble Cay, Torres Strait: results and conclusions from a comprehensive survey in August-September 2014. 'Unpublished report to the Department of Environment and Heritage Protection.' (Queensland Government, Brisbane).

Kim, M. and Pressey, B. (2015). Another Australian animal slips away to extinction. *The Conversation* https://theconversation.com/another-australian-animal-slips-away-to-extinction-36203 (Viewed February 13, 2016).

Latch, P. (2008). Recovery Plan for the Bramble Cay Melomys *Melomys rubicola*. 'Report to Department of the Environment, Water, Heritage and the Arts, Canberra.' (Environmental Protection Agency, Brisbane).

Limpus, C. J., Parmenter, C. J. and Watts, C. H. S. (1983). *Melomys rubicola*, an endangered Murid rodent endemic to the Great Barrier Reef of Queensland. *Australian Mammology* 6, 77-79.

Ramsay, E. P. (1877a). Description of a supposed new species of Rock Wallaby from the Palm Islands; on the north-east coast of Australia, proposed to be called *Petrogale assimilis*. *Proceedings of the Linnean Society of New South Wales*, 1, 359-361.

Ramsay, E. P. (1877b). Zoology of the *Chevert*. Mammals. Part 1. *Proceedings of the Linnean Society of New South Wales* 2, 7-19.

Smith, J. M. B. (1994). Patterns of disseminule dispersal by drift in the north-west Coral Sea. *New Zealand Journal of Botany* 32, 453-461.

Thomas, O. (1924). Some new Australasian Muridae. *Annals and Magazine of Natural History* Series 9: 13, 296-299.

Woinarski, J. C., Burbidge, A. A., and Harrison, P. L. (2015a). Ongoing unraveling of a continental fauna: Decline and extinction of Australian mammals since European settlement. *Proceedings of the National Academy of Sciences* 112, 4531-4540.

Woinarski, J. C., Burbidge, A. A. and Harrison, P. L. (2015b). A review of the conservation status of Australian mammals. *Therya* 6: 155-166.

A Study on the Pools of a Granitic Mountain Top at Moonbi, New South Wales

Brian V Timms

Centre for Ecosystem Science, School of Biology, Earth and Environmental Sciences, University of NSW, Kensington, NSW, 2052.

Published on 7 November 2016 at http://escholarship.library.usyd.edu.au/journals/index.php/LIN

Timms, B.V. (2016). A study on the pools of a granitic mountain top at Moonbi, New South Wales. *Proceedings of the Linnean Society of New South Wales* **138**, 61-68.

Flynns Rock in the Moonbi Ranges has many gnammas (rock pools) that have formed by rock solution and which fill in heavy summer rains and remain inundated for much of the year. The two largest pools support 41 taxa of invertebrates, with the smaller pools less speciose. A rehabilitated gnamma was colonized rapidly by local species. The flora and fauna are comprised almost entirely of widespread eurytopic species dominated by insects, with most typical gnamma genera absent, though *Isoetes*, *Glossostigma*, *Eulimnadia* and *Bennelongia* are represented. Diversity is much influenced by habitat size and to a far lesser extent by isolation.

Manuscript received 19 April 2016, accepted for publication 23 August 2016.

Key Words: aquatic invertebrates, aquatic plants, colonization, pan gnammas, phenology.

INTRODUCTION

Mountain tops almost universally lack standing water, but should they be of granite and flat or domed then weathering pits (or pan gnammas) may form and hold water for weeks or months. Examples abound on the granitic inselbergs of southwestern Western Australia (Pinder et al., 2000) and northwestern Eyre Peninsula (Timms, 2015), on some granitic mountains in the Granite Belt of southern Queensland/northern New South Wales (Webb and Bell, 1979) and on sandstones of Uluru, Australia's iconic inselberg (Timms, 2016a). An instructive example, known locally as Flynns Rock, occurs near Moonbi, New South Wales at the southern edge of the New England granitic massif.

Pan gnammas have been well studied in southwestern Western Australia (Bayly, 1982,1997; Pinder et al., 2000; Jocque et al., 2007; Timms 2012a, 2012b, 2014; Brendonck et al., 2015) revealing a high diversity of invertebrates by world standards (Jocqué et al., 2010; Brendonck et al., 2016) and the influence of major factors such as habitat size and hydrological regime on community structure (Vanschoenwinkel et al., 2009). Yet in pan gnammas in the sandstones of the Sydney basin of eastern Australia invertebrate communities are simply structured though some fauna have some similar adaptations to those of the harsh gnamma environment (Bishop, 1974). The question arises, do granitic gnammas in eastern Australia share this low diversity? Studies on central Victorian granitic gnammas suggest diversity is lower than in Western Australia, but higher than in the Sydney sandstone pools (author, unpublished). Flynns Rock near Moonbi presents another site, though limited in scope and somewhat isolated.

It is the aim of this study to document the pools on this mountaintop by mapping the gnammas and environs, explaining their origin, examining the flora and fauna, and noting their adaptations for living in such an unusual habitat.

THE STUDY SITE AND METHODS

Flynns Rock (Fig 1) is a rectangular block of granite about 35m long by 22 m wide and averaging about 8m above the surrounding mountain slopes. The rock surface slopes 7m north to south and has about 15 enclosed hollows, six of which regularly contain exposed water (labelled 1-6 on Fig. 1). The remainder are filled in with sediment and vegetation, though Nos. 5 and 6 are partially infilled and two (Nos. 7 and 8) were cleaned out during the study (Fig. 1). The dimensions of the six main pools are given in Table 1; conveniently for study, these comprise three pairs of pools, two large, two small and two very small.

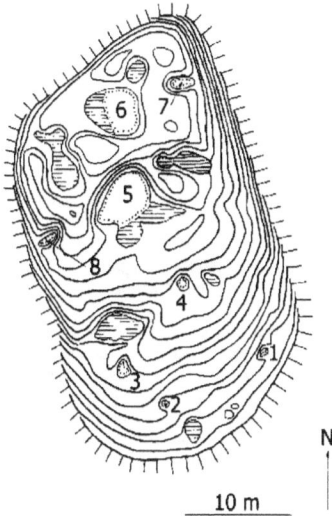

Fig. 1 Map of Flynns Rock. Contour interval 50 cms. Active gnammas shown with dotted edges, infilled ones with horizontal bars. Bars at the rock edge indicate steep slopes.

Pools 7 and 8 were prepared as colonization sites, but studies on pool 8 were abandoned as it has crack near the floor which means it rarely retains water. The rock was mapped in February 2016 using a DJI Phantom 3 professional drone and Agisoft photoscan software.

The rock was visited 14 times over the two year study period (March 2014 to February 2016), generally at about 1 month intervals when the pools contained water, December/March to about September/October. Conductivity was measured with an ADWA AD332 meter and turbidity with a Secchi disc tube calibrated in Nephelometric Turbidity Units (NTU). This tube is not accurate at very low turbidities as values less than five NTU are noted as such rather than a lower figure. Depth (z) as determined with a stiff tape measure and when a pool was overflowing, its length and width (to give the average d) measured and volume calculated. Each was assumed to be saucer-shaped so the formula ($V = (\pi/2) xr^2$) for parabolic shapes was used. Catchments of each pool were independent.

Pool 7 was cleaned out in September 2014 and held water from December 2014 to September 2015 and again in January and February 2016.

Rainfall data were supplied by the Bureau of Meteorology, Station 055321 Mulla Crossing, 12 km south of the study site, with data for November 2014 added from station 055320 Lumbri 15 km away to fill in a gap. The private rain gauge of Warwick Schofield 1 km away from the mountain, but not always read regularly, suggest the Mulla Crossing values used are 5-10% lower than the mountain receives.

Meiofauna was caught in a zooplankton net (opening 10 cm by 8 cm, length 50 cm and mesh 159 µm) with the bottom stirred a little to catch epibenthic species. Macroinvertebrates were caught with a pond net of 1 mm mesh in the two large pools and with 12 cm household sieve of 1 mm mesh in the smaller pools and when the large pools were very shallow. It was difficult to thoroughly clean the nets after each pool as not enough clean water could be carried up to the rock. Pool 7 was always sampled first as a ploy to avoid introductions with possibly contaminated nets. On each sampling occasion, the zooplankton net was

Table 1 Physicochemical features of the pools

Pool	dimensions in cms	maximum depth in cms	mean depth in cms	full volume in litres	Conductivity µS/cm ±SE	Turbidity NTU ± SE
1	125 x 90	6	4.5	27	37.8 ± 7.5	62.4 ± 17.5
2	100 x 80	6	4.4	19	48.6 ± 22.7	50.2 ± 16.1
3	190 x 150	12	9.5	136	91.2 ± 29.3	14.0 ± 3.4
4	170 x 120	9	7.6	74	47.6 ± 9.0	16.3 ± 2.4
5	660 x 460	23	16.2	2832	59.9 ± 8.4	8.7 ± 1.3
6	550 x 500	19	13.5	2056	68.5 ± 11.8	12.3 ± 1.7
7	280 x 180	12	9.3	249	28.8 ± 5.5	29.9 ± 16.3

Table 2 List of plant species found on Flynns Rock

Family	Genus and species	Family	Genus and species
Amaranthaceae	*Alternanthera denticulata*	Moraceae	*Ficus obliqua*
Chenopodiaceae	*Dysphania carinata*	Oxalidaceae	*Oxalis chnoodes*
Chenopodiaceae	*Dysphania pumilio*	Phrymaceae	*Glossostigma elatinoides*
Crassulaceae	*Crassula helmsii*	Poaceae	*Capillipedium spicigerum*
Crassulaceae	*Sedum acre*	Poaceae	*Eragrostris brownii*
Cyperaceae	*Cyperus polystachyos*	Portulacaceae	*Calandrinia eremaea*
Cyperaceae	*Fimbristylis dichotoma*	Pteridaceae	*Cheilanthes distans*
Geraniaceae	*Geranium solanderi*	Pteridaceae	*Cheilanthes sieberi*
Isoetaceae	*Isoetes muelleri*	Ranunculaceae	*Ranunculus inundatus*
Mackinlayaceae	*Xanthosia pilosa*		

used for 1 minute and the macroinvertebrate apparatus for 2 to 3 minutes, depending on pool size. Doubling the sampling time did not add further species. The whole meiofauna collection was preserved in alcohol for later study, but macroinvertebrate collections were sorted alive in a white tray, with representative specimens retained preserved in alcohol for study and the remainder returned alive to the pools. Abundances were estimated on a log scale. I did not have a licence to study tadpoles so the few caught were returned to the pools alive.

RESULTS

The two largest pools contained water April to September in 2014 and December to September in 2014-15 and again January onwards on 2016. The two small pools had a similar hydroperiod but starting earlier in March in 2014, while the two very small pools lacked water both at the beginning and end of the study (ie dry in March 2014 and in 2016). The pools were full only a few times and very shallow mainly in early September; on average they were about three-quarters full (Table 1) so that pool volumes were often about three-quarters those listed in Table 1. Usually each had at least small areas of open water, thus facilitating zooplankton collection, though there were large open areas in the two large pools until macrophytes grew by about May.

Conductivities were always low, averaging less than 100 µS/cm, and often reading < 25 µS/cm when full (Table 1). When water levels were low values up to 344 µS/cm were recorded, but no relationship was noted between pool size/volume and conductivity. Also when pools 1 and 2 were low, turbidities were high (100-200 NTU), but otherwise there was no relationship between pool volume and turbidity, (r

= -0.591, not significant) though the two very small pools were highly turbid and the largest ones clear (Table 1).

Some 19 species of vascular plants were found on the rock (Table 2), but most of these were in the grassed infilled depressions or growing on the shallower, irregularly inundated parts of pools 5 and 6. In the regularly inundated parts of these pools, plus the edges of pools 3 and 4, *Isoetes muelleri* was dominant and persistent. Some *Glossostigma elatinoides* grew patchily in all four pools while *Alternanthera denticulata* was common in the more amphibious parts of pools 5 and 6. The filamentous algae *Oedogonium* sp. and *Zygnema* sp. tended to smother the *Isoetes* by midyear. Pools 1 and 2 lacked plants.

Altogether 41 taxa of invertebrates were found in the six pools, comprising 10 crustaceans and 23 insects (Table 3). Pools 1 and 2, the very small pools lacking vegetation, had the fewest species with just four recorded, the small pools 3 and 4 had 30 taxa and the two large pools had 34 taxa (Table 3). Momentary species richness averaged 2.5 in pools 1 and 2, 5.4 in pools 3 and 4 and 10.2 in pools 5 and 6. Seasonal peaks in species richness varied between years but generally occurred in April/May and minima often at the beginning or end of a season (Fig. 2).

Dipteran larvae dominated in very small pools, and even when dry, the chironomid *Paraborniella tonnoiri* and the ceratopogonid *Dasyhelea* sp. could be extracted from the sediment by adding water. These two species also dominated in pools 3 and 4 with the unidentified brown planarian being common also. Most of the other species occurred spasmodically and often recorded as a single specimen. The large pools 5 and 6 had a variety of species common from time to time, including the brown planaraian, the clam shrimp *Eulimnadia australiensis*, two cladocerans, the ostracod

Proc. Linn. Soc. N.S.W., 138, 2016

63

Table 3 Number of occurrences of invertebrate species in the pools of Flynns Rock, Moonbi

Higher rank	Species	Pool 1	Pool 2	Pool 3	Pool 4	Pool 5	Pool 6	Pool 7
Platyhelminthes	unidentified pink/grey planarian			7	7	9	10	1
	unidentified clear planarian			2		2	2	1
Nematoda	unidentified nematodes	1		5	4	3	2	1
Rotifera	Asplanchna sp.							
	Keratella sp.							
Branchiopoda	Eulimnadia australiensis			3	1	4	4	1
	Daphnia carinata				3			
	Armatalona imitatoria			4	4	5	6	3
	Ephemeroporus sp. (barrosi group)			1		7	4	
	Bennelongia n. sp.					10	9	2
Ostracoda	Cypretta sp.						1	
	Cyprinotus sp.			1		2		
	Ilyodromus sp			2	1	1		
	Sarscypridopsis sp.			1		2	1	2
Copepoda	Mesocyclops notius				1	2	5	1
Odonata	Austrolestes leda					5	4	
	Austrolestes heterostricta			2		1		
	Hemicordulia tau							2
Ephemeroptera	Cloeon sp.			1	1	5	3	2
Hemiptera	Agraptocorixa parvipunctata			4	1	6	7	1
	Micronecta sp.			3	3	12	11	3
	Anisops gratis				1	7	9	3
	Anisops thienemanni			1	1	5	4	1
	Anisops spp.					7	4	
	Enithares sp.							
Coleoptera	Allodessus larvae					6	4	
	Berosus larvae			3	2	5	1	1
	Sternopriscus larvae			2	1	1	2	3
	a hydrophilid larvae			2		1		
	Allodessus histrigatus						1	1
	Antiporus gilberti						1	
	Berosus sp 1			1		2	3	
	Berosus sp 2			1		3	2	
	Sternopriscus multimaculatus			3	4	1	1	
Diptera	Aedes alboannulatus sensu lato	5	6	4		2		
	Paraborniella tonnoiri	10	11		14	8	9	7
	Procladius sp.				1			
	Dasyhelea sp.	10	11	7	10	2	3	6
	Eristalis sp.					2	2	
	unidentified maggot							1
Arachnida	unidentified hydrocarinid mite			1				
Mollusca	Ferrissia sp.						1	2

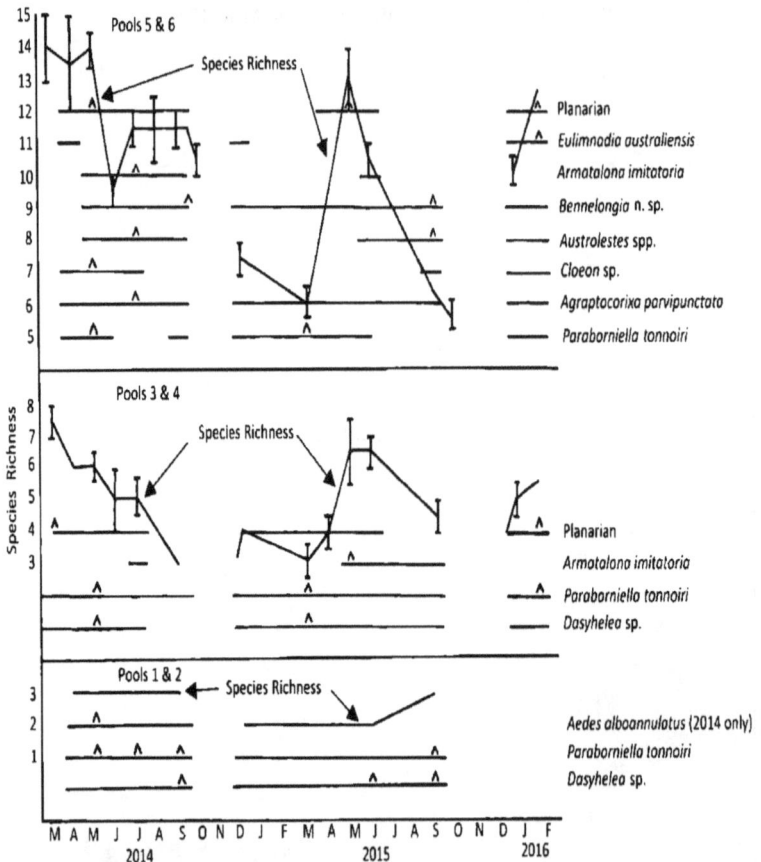

Fig. 2. Invertebrate phenology at Flynns Rock. Species richness curves based on average of two values for the paired pools; variability shown by vertical bars, but on many dates (all dates for pools 1 & 2) both values the same. Seasonal distribution of major species are indicated by horizontal lines, with inverted v's showing peaks in abundance.

Bennelongia n. sp., the odonatan *Austrolestes leda*, the mayfly *Cloeon* sp., the hemipterans *Agraptocorixa parvipunctata, Micronecta* sp. and *Anisops* spp. plus *Paraborniella tonnoiri*.

The dominant species had very different phenologies. *Eulimnadia australiensis* only appeared briefly at the first filling each season (Fig 2). *Cloeon* sp. developed early in each filling cycle and variation in specimen sizes suggested at least 2 generations. *Austrolestes leda* and *Hemicordulia tau* were much slower developers, only appearing numerous later

in the season and there was only one cohort per year (Fig 2). *Anisops* spp. were caught early in each season, bred and then persisted. The two years were not exactly the same in occurrences and abundances, the second year, 2015, had fewer cladocerans, *Cloeon* sp, odonates and no mosquitoes.

The first colonizer in Pool 7 was the chironomid *Paraborniella tonnoiri* which was present in small numbers in the December filling. It was joined by *Dasyhelea* sp. by March and both were abundant by May. A few hemipterans appeared in

Proc. Linn. Soc. N.S.W., 138, 2016

65

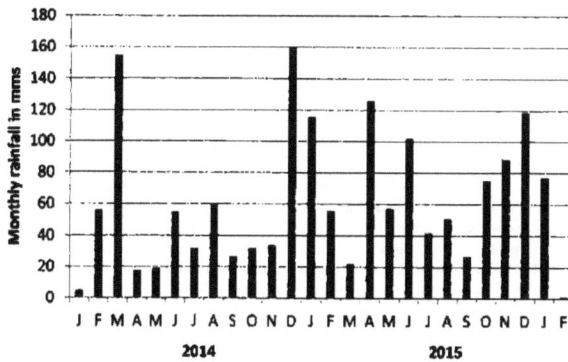

Fig. 3. Rainfall nearby to Flynns Rock.

March and the brown flatworm had arrived by May. The first crustacean to be present was the claoderan *Armatalona* by June followed by the ostracod *Bennelongia* sp. in September. The first *Cloeon* sp appeared also in September. The new filling in January 2016 added the ostracod *Candonocypris* and the clam shrimp *Eulimnadia australiensis* and a sparse population of hemipterans and some dytiscid larvae. The clam shrimp had disappeared again by February (but present still in pools 3-6).

DISCUSSION

The gnammas of Flynns Rock are of two types: pools 1-4 are simple pan gnammas and 5-8 armchair pans (Timms and Rankin, 2016). Pools 1-4 have slightly sloping shore profiles indicative of weathering along surface exfoliation laminations (Twidale and Corbin, 1963) while pools 5-8 have a characteristic exponentially-curved shore profile indicative of water layer weathering. Such armchair pans begin as shallow pans where lamination-controlled weathering predominates but as they incise water layer weathering dominates and the back (and sides) are lowered by subaerial weathering (Timms and Rankin, 2016). Incision is a relatively large 2.5 m in pool 5 (Fig. 1), indicating a much older age for this pool than the others, especially pools 1-4.

The large difference in volumes of the six main pools mean that the small shallow pools, especially 1 and 2, have shorter hydroperiods and it is suspected they dried occasionally between visits.

Pools 5 and 6 once filled early in the season remained inundated for 8-10 months. Generally these pools filled in a summer month with >100 mm rainfall (December to March) then the monthly totals of 20-50 mm coupled with lower evaporation in winter, maintained water for much of the year, till increasing evaporation in spring dried the pools in September/October (Fig 3). This is a different filling-drying cycle from that for the gnammas in the mediterranean climate of southwest Western Australia and Eyre Peninsula, South Australia where pools fill in late autumn or early winter, retain water during the winter and dry in spring (Timms, 2012a, 2014).

Conductivities are low compared to those of these western gnammas, this is no doubt influenced by the relative greater salt load in the rain feeding the western gnammas (Hutton and Leslie, 1958; Timms and Rankin, 2016) and by the overflowing of the Moonbi gnammas generally at least once a year during the heavy rainfalls of summer. Turbidities are higher however, probably due to the denser plant populations and higher organic matter load, though this has not been quantified. Aquatic plant growth, particularly of *Isoetes muelleri* (a quillwort) in pools 5 and 6, is luxuriant compared to that in western gnammas (Timms, 2014). *Isoetes* is widespread in gnammas and a characteristic genus in many (Hopper et al., 1997; author, unpublished).

Diversity and community composition of invertebrates are very different from those in these western gnammas (Bayly, 1982,1997; Pinder et al., 2000; Jocque et al., 2007; Timms 2012a, 2012b, 2014; Brendonck et al., 2015). A typical rock outcrop in southwest WA may have 60-70 species (Jocque et al., 2007) and an individual pool 30-40 species (Timms, 2012a), all dominated by crustaceans with many regional endemics (Pinder et al., 2000). Pools 5 and 6 approach this diversity, but crustaceans are few and there is only one possible endemic. The fauna of pools 1-4 is restricted by their small size (Vanschoenwinkel et al., 2009), but again the comparative lack of crustaceans is the salient feature.

Almost all the invertebrates of the Moonbi gnammas are eurytopic species (ie widespread and tolerant). The only possible exception is the new *Bennelongia* sp. and maybe the planarians when they are identified. While various clam shrimps are often endemic in gnammas (Timms, 2016a) the species (*Eulimnadia australiensis*) in Flynn Rock gnammas is widespread in northeast Australia and moreover lives in a variety of habitats (Timms, 2016a). The few cladocerans and other ostracods present are also widespread and not restricted to gnammas, a contrast to a significant component of the fauna of gnammas of Western Australia (Pinder et al., 2000).

Only the chironomid *Paraborniella tonnoiri* and the ceratopogonid *Dasyhelea* sp. have cryptobiotic adaptations to survive in the temporary environments of these gnammas. As such, they are well suited to the precarious fluctuating habitat provided by pools 1-4. The crustaceans present are preadapted for temporary environments in that they lay eggs capable of surviving the dry times. Most of the insect inhabitants take advantage of the temporary presence of water which generally lasts long enough for many to breed successfully, though perhaps isolation of the pools on a mountain top may restrict dispersal as it apparently did for mosquitoes in 2015. Colonization of the new pool was restricted to fauna already in nearby pools on the rock, again suggesting the isolated mountain top position may be restrictive. Though tadpoles were encountered from time to time, they were not regular and predictable faunal component, suggesting breeding frogs could be restricted by the rock's high steep sides and isolation from other waters.

The conclusion is that these gnammas, while physically similar to many elsewhere, support a generalised fauna with few species characteristic of gnammas. For smaller temporary waters a study of just the crustaceans of 41 pools in southeastern Victoria yielded an average of 9.3 species per pool (Morton and Bayly, 1977), well in excess of those in the Moonbi pools. This low diversity of the Moonbi pools is largely due to their small size and also to the lack of long term climatic variation thought to contribute to the relatively high faunal diversity in southwestern Australian gnammas (Pinder et al.,

2000). The low diversity is also partially due to their isolation on a mountain top, with dispersal from rock pools nearby often essential to maintain local diversity (Vanschoenwinkel et al., 2013). However, at Moonbi this is thought to be of minor influence, comparable to very low faunal diversity in desert gnammas of southeastern Western Australia, where lack of similar pools in the greater district and their very small size to receive colonizers, impose severe restrictions (Bayly et al., 2011). On the other hand, compared to the diversity in Sydney basin gnammas (Bishop 1974) and to gnammas in the Granite belt of Southeast Queensland (author unpublished data) the comparatively larger Moonbi pools are speciose, probably because they are vegetated (ie. more complex habitat structure as well as larger habitat size) (Vanschoenwinkel et al., 2009). However, both these gnamma groups have a specialised endemic limnadiid clam shrimp (Timms, 2016b), a specialisation lacking at Moonbi.

ACKNOWLEDGEMENTS

I thank the Goodfellow family for access to their special mountaintop rock and to Warwick and Margie Schofield and Barbara Moritz for leading the initial scramble to find the rock. I had numerous field assistants, including John Vosper and Terry Annable who came many times, for which I am grateful. Terry took a special interest in the plants and I thank him for his plant list. I am grateful to the following taxonomists for their identifications: Peri Coleman (algae), Stuart Halse (ostracods), Chris Mudden (*Paraborniella tonnoiri*), Russ Shiel (cladocera, *Mesocyclops*) Chris Watts (*Paracymus pygmaeus*), and Craig Williams (*Aedes alboannulatus*). Thanks to Justin McCann for training me to use the drone and for map data analysis. Finally I am indebted to Warwick Schofield and to two anonymous reviewers for helpful comments on the manuscript.

REFERENCES

Bayly, I.A.E. (1982). Invertebrates of the temporary waters on granite outcrops in southern Western Australia. *Australian Journal of Marine and Freshwater Research* 33: 599-606.

Bayly, I.A.E. (1997). Invertebrates of temporary waters in gnammas on granite outcrops in Western Australia. *Journal of the Royal Society of Western Australia* 80: 167-172.

Bayly, I.A.E., Halse, S.A. and Timms, B.V. (2011). Aquatic invertebrates of rockholes in the south-east of Western Australia. *Journal of the Royal Society of Western Australia* 94: 549-555.

Bishop, J. A., (1974). The fauna of temporary rain pools in eastern New South Wales. *Hydrobiologia* 44: 319-323.

Brendonck. L., Jocqué, M., Tuytens, K, Timms, B.V. and Vanschoenwinkel, B. (2015). Hydrological stability drives both local and regional diversity patterns in rock pool communities *Oikos* 124: 741-761.

Brendonck. L., Lanfranco. S., Timms, B. and Vanschoenwinkel, B. (2016). Invertebrates in Rock Pools. In: *'Invertebrates in Freshwater Wetlands'* (eds Batze, D., & Boix, D.) pp 25-53. (Springer: Switzerland.)

Hopper, S.D, Brown, A.P. and Marchant, N.G. (1997). Plants of Western Australian granite outcrops. *Journal of the Royal Society of Western Australia* 80: 141-158.

Hutton, J.T. & Leslie, T.I. (1958). Accession of non-nitrogenous ions dissolved in rainwater to soils in Victoria. *Australian Journal of Agricultural Research* 9: 492-507.

Jocqué, M, Timms, B.V. and Brendonck, L. (2007). A contribution on the biodiversity and conservation of the freshwater fauna of rocky outcrops in the central Wheatbelt of Western Australia. *Journal of the Royal Society of Western Australia* 90: 137-142.

Jocqué, M.. Vanshoenwinkel, B., and Brendonck, L. (2010). Freshwater rock pools: a review of habitat characteristics, faunal diversity and conservation value. *Freshwater Biology* 55: 1587-1602.

Morton. D.W. and Bayly, I.A.E. (1977). Studies on the ecology of some temporary freshwater pools in Victoria with special reference to microcrustaceans. *Australian Journal of Freshwater and Marine Research* 28(4): 439-454.

Pinder, A.. Halse, S.A. Shiel, R.J. and McRae, J. M. (2000). Granite outcrop pools in south-western Australia: foci of diversification and refugia for aquatic invertebrates. *Journal of the Royal Society of Western Australia* 83: 149-161.

Timms, B.V. (2012a). Seasonal study of aquatic invertebrates in five sets of latitudinally separated gnammas in southern Western Australia. *Journal of the Royal Society of Western Australia* 95: 13-28.

Timms, B.V. (2012b). Influence of climatic gradients on metacommunities of aquatic invertebrates on granite outcrops in southern Western Australia. *Journal of the Royal Society of Western Australia* 95: 125-135.

Timms, B.V. (2014). Community ecology of aquatic invertebrates in gnammas (rock-holes) of north-western Eyre Peninsula, South Australia. *Transactions of the Royal Society of South Australia* 138: 147-160.

Timms, B.V. (2016a). A partial revision of the Australian *Eulimnadia* Packard, 1874 (Branchiopoda: Spinicaudata: Limnadiidae). *Zootaxa* 4066 (4): 351-389.

Timms, B.V. (2016b). A review of the Australian endemic clam shrimp *Paralimnadia* (Branchiopoda: Spinicaudata: Limnadiidae). *Zootaxa* 4161(4): 451-508.

Timms, B.V. and Rankin, C. (2016). The geomorphology of gnammas (weathering pits) of northwestern Eyre Peninsula, South Australia: typology; influence of haloclasty and origins. *Transactions of the Royal Society of South Australia* 140: 28-45.

Twidale, C.R. and Corbin, E.M.(1963). Gnammas *Revue de Geomorphology dynamique* 14:1-20.

Vanschoenwinkel, B., Hulsman, A., de Roeck, E., de Vries, C., Seaman, M. and Brendonck, L. (2009). Community structure in temporary freshwater pools: disentangling the effects of habitat size and hydroregime. *Freshwater Biology* 54: 1487-1500.

Vanschoenwinkel, B., Buschle, F.T and Brendonck, L. (2013). Disturbance regime alters the impact of dispersal on alpha and beta diversity in a natural metacommunity. *Ecology* 94: 2547-2557.

The Beach Stone-Curlew (*Esacus magnisrostris*) in the Sydney Basin and South East Corner Bioregions of New South Wales

MATTHEW MO

NSW Department of Primary Industries, Elizabeth Macarthur Agricultural Institute, Woodbridge Road, Menangle, New South Wales 2568 (matthew.mo@dpi.nsw.gov.au)

Published on 16 December 2016 at http://escholarship.library.usyd.edu.au/journals/index.php/LI

Mo, M. (2016). The beach stone-curlew (*Esacus magnisrostris*) in the Sydney Basin and South East Corner bioregions of New South Wales. *Proceedings of the Linnean Society of New South Wales* 138, 69-81.

The beach stone-curlew (*Esacus magnirostris*) has only been resident in New South Wales since the 1970's. Here, records of the beach stone-curlew from the Sydney Basin and South East Corner bioregions were analysed. The earliest record was an individual sighted on The Entrance in 1959, with no subsequent records until 1978. The majority of records in the study area occurred in the last two decades. Some records are isolated; however others indicate that a bird may have stayed in an area for up to months at a time. Examples include individuals seen at the Long Reef Aquatic Reserve, Botany Bay/Kurnell Peninsula, the Royal National Park, the Shoalhaven district and Merimbula. Based on the timing proximity of some records in 2015, records from four locations in Sydney in the space of two months may have been the same individual. Few sightings of more than one bird were recorded. Merimbula was the southern extremity of beach stone-curlews in the study area. The known breeding population in the state was restricted to the North Coast bioregion prior to the recent observations of a breeding pair in Port Stephens.

Manuscript received 21 August 2016, accepted for publication 29 November 2016.

KEY WORDS: dispersion, distribution, movements, threatened species, vagrant

INTRODUCTION

The beach stone-curlew (*Esacus magnisrostris*) (Fig. 1) is listed as critically endangered in New South Wales (NSW) under the Threatened Species Conservation Act 1995 (TSC Act). This large shorebird, usually seen singly or in pairs, is found from Southeast Asia to northern Australia (Blakers *et al.* 1984; Freeman 2003; Trainor 2005), occurring exclusively in coastal littoral habitats, such as river mouths, mudflats, sandbars and beaches (Clancy 1986; Garnett 1992). In Australia, the population was reported as stable by Garnett and Crowley (2000), however, authors have been divided over whether it is declining (Garnett 1993; Watkins 1993) or increasing in NSW (Smith 1991).

Until recently, the known breeding population in NSW was restricted to the North Coast bioregion (Clancy and Christiansen 1980; Clancy 1986; Hole *et al.* 2001). Beach stone-curlews are seldom located south of the Manning River, which was the southern extremity of Rohweder's (2003) study on the NSW population. They are considered a rare vagrant to the Sydney Basin bioregion and more southerly locations (Hoskin *et al.* 1991; Schulz and Ransom 2010). The occurrence of beach stone-curlews in these parts often draws the attention of bird observers keen to experience the rarity.

In this paper, I review the occurrence of the beach stone-curlew in the Sydney Basin and South East Corner bioregions of NSW. This work is important in understanding the ecology of this threatened species in the southern parts of its distribution.

METHODS

Records of beach stone-curlews sighted in the Sydney Basin and South East Corner bioregions of New South Wales (study area; Fig. 2) were collated. This paper follows the bioregion definitions set out by the NSW National Parks and Wildlife Service (2003). The Sydney Basin bioregion is the region encompassed by the Hunter region in the north to the Shoalhaven region in the south, and the South East Corner bioregion encompasses the remainder of the NSW coastline south of the Sydney Basin bioregion.

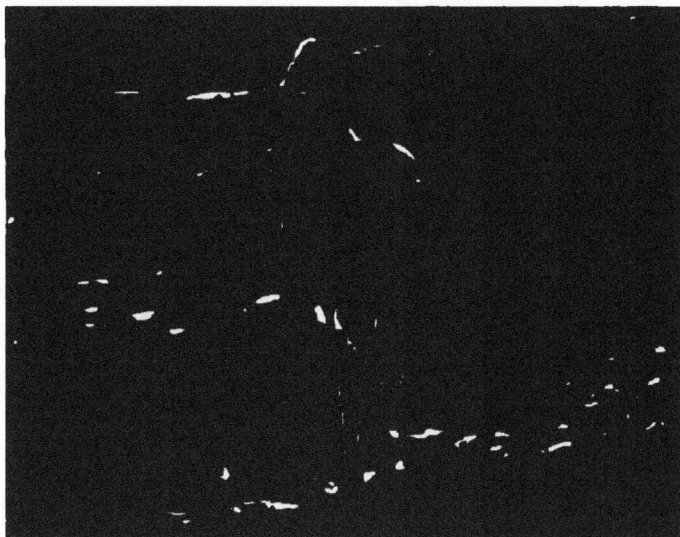

Figure 1. A beach stone-curlew (*Esacus neglectus*) sighted at Merimbula, South East Corner of New South Wales, in March 2016. Photo, John Bundock.

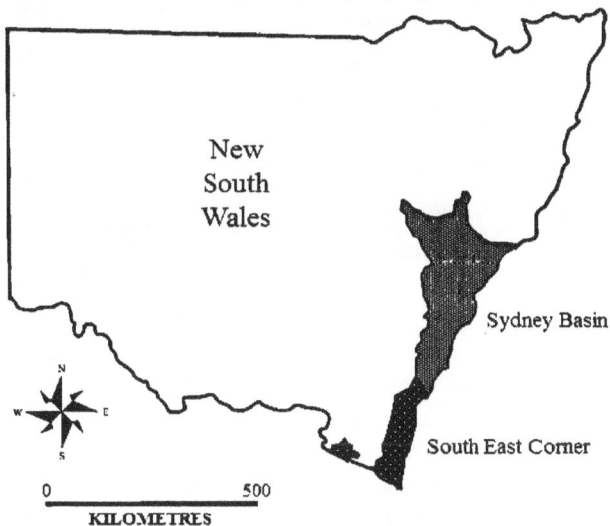

Figure 2. The Sydney Basin and South East Corner bioregions in the context of New South Wales.

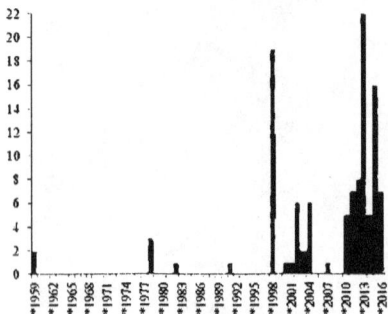

Figure 3. Beach stone-curlew (*Esacus neglectus*) records in the Sydney Basin and South East Corner bioregions of New South Wales.

Information was sought from databases such as the Atlas of Living Australia (ALA 2016), the Atlas of NSW Wildlife (OEH 2016), the Cumberland Birds Observers Club (CBOC) database, the Eurobodalla Natural History Society (ENHS) database, the Far South Coast Birdwatchers Inc. (FSCB) database and Birdata. The latter is maintained by BirdLife Australia, which functions as a web portal for members to submit bird sighting information. In addition, this organisation (formerly the Royal Australasian Ornithologists Union, then Birds Australia) also administers the Historical Bird Atlas, which is a collection of records from between 1629 and 1976 sourced from museum collections, published literature and unpublished sources such as personal notebooks. The organisation has conducted two nation-wide atlas surveys, the first of which (Blakers *et al.* 1984) included records relevant to this study.

Additional information was retrieved from online posts by bird observers to websites such as Birding-Aus (2016) and Eremaea Birdlines (2016), as well as personal communications.

RESULTS

Records of the beach stone-curlew in the study area have accumulated mostly in the last two decades (Fig. 3). The increase in recent records may be due to the increasing number of observers and the ease of reporting observations on internet-based databases. The first record was in 1959 (discussed below), after which there was a 19-year period until subsequent records (Table 1). Figure 4 shows some

of the locations the beach stone-curlew has been recorded within the study area.

Central Coast

The earliest record on the Central Coast was a single beach stone-curlew sighted at The Entrance North on 27 December 1959 (Blakers *et al.* 1984; Table 1). On the same day, it was also recorded at Norah Head, 8 km north (Wilson 1961; Stringfellow 1962). There were no subsequent sightings until 7 September 1991 when one individual was seen at Tuggerah Lake (Eremaea Birdlines 2016).

Another sporadic record occurred at Stockton sandspit in the Hunter Wetlands National Park on 5 December 2002, with a subsequent record at this locality on 3 October 2015. In 2013, the Entrance was visited by at least one beach stone-curlew, sighted on 12 October and 22 November (Eremaea Birdlines 2016).

Port Stephens

Since 2011, a pair of beach stone-curlews (presumably the same pair) have been reliably viewed at Soldiers Point and Dowardee Island, Port Stephens (Morris *et al.* 2011; Murray 2013, 2014; Birding-Aus 2016). They appear to reside on Dowardee Island, where they have been observed at least five times (Morris *et al.* 2011; Eremaea Birdlines 2016), flying over to Soldiers Point, mainly to feed at low-tide, where most sightings were recorded (Murray 2013; ALA 2016). On the mainland, the beach stone-curlews drink and bathe from a stormwater outlet. When disturbed by people, they flew back to Dowardee Island.

The pair had bred successfully for at least four summers (HBOC 2015). Each year, the beach stone-curlews become absent from Soldiers Point from October to February, reappearing on the mainland in late summer with a near-independent fledgling (Murray 2013). Breeding is thought to occur on Dowardee Island (Murray 2013), although no actual signs of nesting have been reported to date.

On 20 March 2015, one individual was observed at Lemon Tree Passage (Eremaea Birdlines 2016), which is presumably one of the pair from Dowardee Island. The distance between these two locations is only 3 km.

Northern Sydney

There are records of a beach stone-curlew at Dee Why Lagoon for 3 November 2010 (Eco Logical Australia 2011; Birding-Aus 2016). Three years later, an individual beach stone-curlew was seen in

Proc. Linn. Soc. N.S.W., 138, 2016

71

BEACH STONE-CURLEWS IN SOUTH EASTERN NEW SOUTH WALES

Table 1. Records of the beach stone-curlew (*Esacus neglectus*) in the Sydney Basin and South East Corner bioregions.
ALA = Atlas of Living Australia, CBOC = Cumberland Bird Observers Club, EB = Eremaea Birdlines, ENHS = Eurobodalla Natural History Society, FSCB = Far South Coast Birdwatchers Inc., HBA = Historical Bird Atlas, OEH = Office of Environment and Heritage, NSW Atlas of Wildlife

Location	Date	Source
Central Coast		
The Entrance North	27 Dec 1959	Blakers *et al.* 1984
Norah Head	27 Dec 1959	Wilson 1961, Stringfellow 1962
Tuggerah Lake	7 Sep 1991	EB
Stockton Sandspit	5 Dec 2002	Birdata
The Entrance	12 Oct 2013	EB
The Entrance	22 Nov 2013	EB
Stockton Sandspit	3 Oct 2015	EB
Port Stephens		
Dowardee Island	20 Jan 2011	EB; HBOC
Soldiers Point	22 May 2011	EB; HBOC
Dowardee Island	24 May 2011	Birdata
Dowardee Island	2 Aug 2011	Birdata
Dowardee Island	11 Aug 2011	EB
Dowardee Island	14 Aug 2011	Morris *et al.* 2011
Soldiers Point	26 Mar 2012	Birdata
Soldiers Point	3 Jul 2012	Birdata
Soldiers Point	16 Aug 2012	Birdata
Soldiers Point	18 Aug 2012	EB
Soldiers Point	26 Aug 2012	Birdata
Soldiers Point	28 Aug 2012	EB
Soldiers Point	12 Oct 2013	ALA
Soldiers Point	29 Mar 2014	EB
Lemon Tree Passage	20 Mar 2015	EB
Northern Sydney		
Brooklyn	n/d	HBA
Dee Why Lagoon	3 Nov 2010	CBOC, Birding-Aus; Eco Logical Australia 2011
Manly	14 Oct 2013	OEH
Long Reef Aquatic Reserve	23 Nov 2013	CBOC, EB
Kissing Point Park, Putney	14 Nov 2015	CBOC, EB
Long Reef Aquatic Reserve	28 Nov 2015	EB
Long Reef Aquatic Reserve	29 Nov 2015	EB
Long Reef Aquatic Reserve	4 Dec 2015	CBOC
Long Reef Aquatic Reserve	5 Dec 2015	CBOC, EB
Long Reef Aquatic Reserve	6 Dec 2015	CBOC, EB
Long Reef Aquatic Reserve	7 Dec 2015	EB
Long Reef Aquatic Reserve	8 Dec 2015	EB
Long Reef Aquatic Reserve	9 Dec 2015	CBOC

72

Long Reef Aquatic Reserve	11 Dec 2015	EB
Long Reef Aquatic Reserve	12 Dec 2015	CBOC

Botany Bay/Kurnell Peninsula

Towra Point Nature Reserve*	1982	OEH
Bonna Point, Kurnell	2 Jun 1998	CBOC, EB
Towra Point Nature Reserve	Dec 2001	Birding-Aus
Metromix Swamp, Kurnell	29 Nov 2003	CBOC
Boat Harbour, Kurnell	20 Dec 2003	CBOC
Towra Point Nature Reserve	27 Jan 2004	CBOC
Boat Harbour, Kurnell	3 Feb 2004	CBOC
Towra Point Nature Reserve	7 Feb 2004	OEH
Boat Harbour, Kurnell	3 Mar 2004	CBOC
Towra Point Nature Reserve	February 2010	OEH 2013
Towra Point Nature Reserve	20 Nov 2010	CBOC, OEH
Towra Point Nature Reserve	Nov 2011	OEH 2013
Taren Point Shorebird Reserve	16 Nov 2015	CBOC, EB
Near Sydney Airport	18 Nov 2015	CBOC

Royal National Park

Era Beach	1998-1999	DECCW 2011
Bundeena	4 Apr 1998	ALA
Bundeena	30 May 1998	CBOC
Bundeena	14 Jun 1998	CBOC, Birding-Aus
Bundeena	21 Jun 1998	CBOC
Bundeena	28 Jun 1998	CBOC
Maianbar and Bundeena	3 Jul 1998	CBOC
Royal National Park	4 Jul 1998	Birding-Aus
Bundeena	5 Jul 1998	CBOC
Bundeena	11 Jul 1998	CBOC
Bundeena	25 Jul 1998	ALA
Bundeena	27 Jul 1998	CBOC
Bundeena	13 Aug 1998	ALA, CBOC
Bundeena	14 Aug 1998	CBOC
Bundeena	27 Nov 1998	ALA, CBOC
Port Hacking	2000	Breen 2007, Birding-Aus, EB
Maianbar and Bundeena	8 Mar 2004	CBOC
Bundeena	12 Mar 2004	CBOC
Deeban Spit, Maianbar	6 Nov 2010	ALA, CBOC, EB
Bundeena	14 Nov 2010	ALA, CBOC

Illawarra

Thirroul	28 Feb 1998	OEH
Windang	30 Jan 2013	Birdata, EB; Cocker 2013
Windang	14 Oct 2014	EB

Shoalhaven

Orient Point	9 Jan 1978	Blakers et al. 1984
Shoalhaven Heads	1 Feb 1978	OEH
Comerong Island Nature Reserve	1 Feb 1978	OEH
Orient Point	4 May 1998	OEH
Shoalhaven Heads	3 Feb 2002	OEH
Comerong Island Nature Reserve	24 Feb 2002	OEH
Orient Point	25 Feb 2002	Birding-Aus
Baileys Island, Gerroa	1 Nov 2007	Birdata
Orient Point	17 Oct 2012	EB
Lake Wollumboola	27 Jan 2013	OEH
Orient Point	5 Feb 2013	EB
Lake Wollumboola	10 Feb 2013	D. Paton, pers. comm
Orient Point	17 Feb 2013	EB
Shoalhaven Heads	27 Feb 2013	EB
Orient Point	1 Jun 2013	EB
Orient Point	17 Jul 2013	OEH
Comerong Island Nature Reserve	1 Dec 2013	EB
Orient Point	1 Jul 2014	OEH
Culburra Beach	15 Jul 2014	OEH
Comerong Island Nature Reserve	3 Nov 2014	EB

South East Corner

Merimbula	1 Sep 1998	Birdata
Durras North, Murramarang National Park	29 Dec 1998	OEH
Moruya	2002	Morgan 2013; ENHS
Mogareeka Inlet	17 Dec 2002	Birdata
Wallaga Lake	2012	Morgan 2013; ENHS
Burrill Lake	2 Jan 2013	EB
Lake Tabourie	21 Oct 2013	OEH
Tuross Head	2 Dec 2013	Morgan 2013; ENHS
Lake Tabourie	10 Dec 2013	OEH
Toragy Point, Moruya Heads	16 Dec 2013	EB
Burrill Lake	20 Dec 2013	EB
Spencer Park, Merimbula	24 Oct 2015	Birding-Aus
Spencer Park, Merimbula	26 Oct 2015	SCRSH 2015
Spencer Park, Merimbula	15 Mar 2016	FSCB
Spencer Park, Merimbula	17 Mar 2016	ALA
Spencer Park, Merimbula	19 Mar 2016	Birding-Aus
Spencer Park, Merimbula	2 Apr 2016	ALA
Spencer Park, Merimbula	25 Apr 2016	Birding-Aus
Spencer Park, Merimbula	6 May 2016	FSCB
Spencer Park, Merimbula	15 May 2016	FSCB

*Reported as a bush stone-curlew, probably erroneously referring to beach stone-curlew

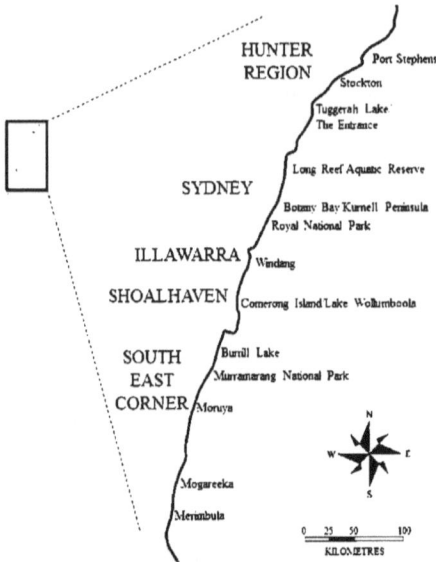

Figure 4. Map of the study area showing some locations where beach stone-curlews *Esacus neglectus* have been recorded.

the area, first at Manly on 14 October 2013 (OEH 2016), then at Long Reef Aquatic Reserve (Fig. 5) on 23 November 2013 (Eremaea Birdlines 2016). Possibly, the same individual was sighted in both records.

An individual was reliably located at Long Reef Aquatic Reserve from 28 November to 12 December 2015. Just prior to these sightings, a beach stone-curlew (possibly the same individual) was also seen along the Parramatta River at Putney on 14 November. The observer at Putney remarked that the bird was "quite relaxed", feeding on crabs within 20 m of the boat ramp (Eremaea Birdlines 2016). These records came after the record at Stockton sandspit.

There is also an undated record from Brooklyn in the Historical Bird Atlas.

Botany Bay/Kurnell Peninsula

In 1982, a bird reported as a bush stone-curlew (*Burhinus grallarius*) was seen in the Towra Point Nature Reserve. Given the habitat where the sighting occurred, the record was probably erroneously referring to a beach stone-curlew (OEH 2013). Since then, a number of irregular sightings have occurred at the Woolooware Shorebird Lagoon, Pelican Point, Towra Spit Island and Towra Beach (Murray and Dessmann 2012; OEH 2013). One beach stone-curlew was

Figure 5. Rock platform and beach at the Long Reef Aquatic Reserve, Collaroy. Photo, M. Mo

Proc. Linn. Soc. N.S.W., 138, 2016

75

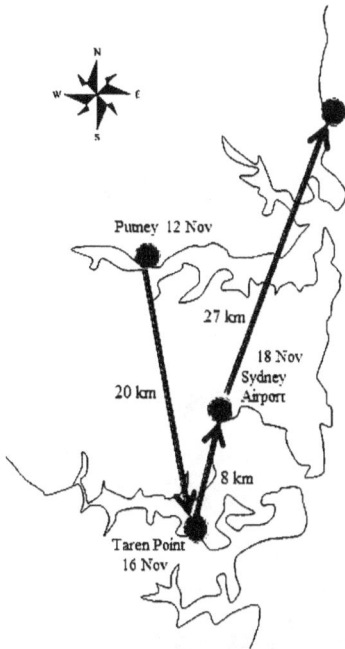

Figure 6. Movements of the beach stone-curlew *Esacus neglectus* in Sydney between November and December 2015, assuming the records were represented by the same individual.

sighted on Bonna Point, Kurnell on 2 June 1998 (Eremaea Birdlines 2016), which coincided with the period of time that an individual was regularly sighted in the northern portion of the Royal National Park (discussed below). There was a further sighting in the Towra Point Nature Reserve in December 2001 (Birding-Aus 2016).

From November 2003 to March 2004, at least one beach stone-curlew was observed at various locations on the Kurnell Peninsula on at least six occasions. The first of these records was made at the Metromix Swamp on 29 November 2003. There were at least three sightings at Boat Harbour on the southern end of the Peninsula in December 2003 and February and March 2004. The remainder of the sightings were recorded at the Towra Point Nature Reserve (OEH 2016). Further sightings in this vicinity followed in

February and November 2010 and November 2011 (OEH 2013, 2016), which may be different individuals given the great length of time between records.

Most recently, one beach stone-curlew was recorded at the Taren Point Shorebird Reserve on 16 November 2015 (Eremaea Birdlines 2016), which represents the most westerly record of this species in Botany Bay. It was not found there the following day (W.A. Hewson, pers. comm). On 18 November, it was seen near Sydney Airport (presumably the same individual). These two records came just after the record at Putney and precede the records from the Long Reef Aquatic Reserve within the same month (mentioned above; Eremaea Birdlines 2016). These records may be the same individual briefly moving between these locations (Figure 6).

Royal National Park

Clarke and Dolby (2014) refer to the beach stone-curlew amongst the shorebird assemblage in the Royal National Park; however Schulz and Magarey (2012) associated it with vagrant status. One individual was observed on the mudflats at Bundeena near the Bonnie Vale camping ground (Fig. 7) on at least 14 occasions between April and November 1998 (ALA 2016; Birding-Aus 2016). It was also seen at least once at Deeban Spit, Maianbar (Fig. 8), which is less than 1 km to the west on the other side of Cabbage Tree Basin. During this time and into 1999, a beach stone-curlew was also sighted further south at Era Beach from 1998 to 1999 (DECCW 2011), possibly the same individual. The following year, there was an isolated sighting of a beach stone-curlew near Bundeena (Breen 2007; Eremaea Birdlines 2016; Birding-Aus 2016).

There was a second occurrence of a beach stone-curlew in Maianbar and Bundeena between 8 and 12 March 2004. These records came just days after the period of time an individual was repeatedly sighted at the Towra Point Nature Reserve and Boat Harbour (OEH 2016), which are only 6 km north, suggesting that the same bird had moved. A third occurrence of an individual in the same locality was recorded on 6 and 14 November 2010 (ALA 2016; Eremaea Birdlines 2016). The beach stone-curlew was sometimes seen foraging in associations with Australian pied oystercatchers (*Haematopus longirostris*), another threatened species. One week after it supposedly moved on, there was the record in the Towra Point Nature Reserve (mentioned above; OEH 2016).

Illawarra and Shoalhaven

The earliest NSW records of beach stone-curlew south of Sydney were an individual sighted at

76

Proc. Linn. Soc. N.S.W., 138, 2016

Figure 7. Mangrove-lined sand flats at the Bonnie Vale camping ground, Bundeena. Photo, M. Mo.

Figure 8. Sand flats at Deeban Spit, Maianbar. Photo, M. Mo

BEACH STONE-CURLEWS IN SOUTH EASTERN NEW SOUTH WALES

Orient Point in the Shoalhaven district on 9 January 1978 (Blakers *et al.* 1984) and an individual that was sighted at Shoalhaven Heads on 1 February 1978 and at the Comerong Island Nature Reserve on the same day (OEH 2016). The first subsequent record for the Illawarra and Shoalhaven districts was from Thirroul on 28 February 1998, preceding another record from Orient Point on 4 May 1998 (OEH 2016). The temporal isolation of these records also suggests the same individual was being located.

A further cluster of sightings occurred in February 2002. Another beach stone-curlew was seen at Shoalhaven Heads on 3 February 2002, with subsequent sightings one day apart at the Comerong Island Nature Reserve and Orient Point toward the end of the month (Birding-Aus 2016; OEH 2016). An isolated sighting of a bird at Baileys Island, Gerroa was recorded on 1 November 2007.

Between October 2012 and November 2014, 14 sightings of the beach stone-curlew were recorded in the Illawarra and Shoalhaven districts, with no more than seven months interval between records. The first record was at Orient Point on 17 October 2012 (Eremaea Birdlines 2016). During its time in the local area, this beach stone-curlew moved mostly between locations from Shoalhaven Heads south to Lake Wollumboola (Eremaea Birdlines 2016; OEH 2016; D. Paton, pers. comm). There was some disturbance from unleashed dogs and fishermen collecting worms from the mudflats (D. Paton, pers. comm). At Lake Wollumboola, it was sometimes close to an Australian pied oystercatcher nest, which prompted it to be mobbed by the brooding adults (OEH 2015). The beach stone-curlew was also seen further north at Windang on at least two occasions (Cocker 2013; Eremaea Birdlines 2016). It was eventually found dead at Orient Point in November 2014 (D. Paton, pers. comm).

South East Corner

The earliest records in the South East Corner bioregion were two records in 1998 that were four months apart. One beach stone-curlew was sighted in Merimbula on 1 September 1998, with the subsequent sighting in Durras North in the Murramarang National Park, 140 km north, on 29 December 1998 (OEH 2016). There were no other known occurrences in the region until a sighting at Moruya in 2002 (Morgan 2013), and one further sighting at Mogareeka Inlet on 17 December 2002.

In 2013, a single beach stone-curlew was located at Wallaga Lake (Morgan 2013). The following year, a cluster of sightings was reported in various locations

in the South East Corner bioregion. One individual was sighted at Burrill Lake, south of Ulladulla, on 2 January 2013. It was foraging close to a nesting colony of little terns (*Sternula albifrons*) and was observed being aggressively mobbed by parent birds. The observer returned to the site later in the day to find it had moved on (Eremaea Birdlines 2016). From October to December 2013, sightings were received from Lake Tabourie twice (OEH 2016), Tuross Head (Morgan 2013) and a second sighting for Burrill Lake (Eremaea Birdlines 2016). On 16 December 2013, a pair of beach stone-curlews was recorded at Toragy Point at Moruya Heads (Eremaea Birdlines 2016). These records did not follow any one direction chronologically, sporadically appearing up and down the coastline.

A beach stone-curlew was reported twice in the same week at Spencer Park, Merimbula in October 2015 (SCRSH 2015; Birding-Aus 2016). Merimbula is the southern extremity of beach stone-curlew sightings in NSW. One bird was first observed on 24 October 2015, but was not located the following day (Birding-Aus 2016). It was however seen at the site by another observer on 26 October (SCRSH 2015).

A second cluster of sightings at Spencer Park occurred between March and May 2016 (ALA 2016; Birding-Aus 2016). A representative of Far South Coast Birdwatchers Inc. warned people through the local newspaper to minimise disturbance to the bird (Anon. 2016). The beach stone-curlew sought refuge in the mangroves behind the beach, moving onto the sand flats to forage (J. Bundock, pers. comm). There was some concern that the influx of holidaymakers over the Easter season could potentially cause the bird to move off. During this time, a second beach stone-curlew joined the first individual (Birding-Aus 2016). The two individuals were first sighted together on 25 April. The last date either bird was recorded was 15 May.

DISCUSSION

The beach stone-curlew has only been resident in northern NSW since the 1970's (Morris *et al.* 1981). It appears to be expanding its distribution southward along the east coast of Australia (Marchant and Higgins 1993). In 1925, an individual seen in Moreton Bay, Queensland, at the mouth of the Brisbane River was the most southerly recorded sighting at the time (Mayo 1925). The first record in NSW was a beach stone-curlew sighted in Tweed Heads in 1930 (Marchant and Higgins 1993), with no subsequent record until 1959, which was the individual sighted at The Entrance (Wilson 1961; Stringfellow 1962),

78

Proc. Linn. Soc. N.S.W., 138, 2016

mentioned in this paper. In 2001, the state population was thought to be 12 individuals (Hole *et al.* 2001).

Movements in the beach stone-curlew are not well known. They probably never disperse far from the coast (Amiet 1957). Pairs can be present in the same location for several years. Examples include the breeding pairs present at Red Rock between 1976 and 1986 (Clancy 1986), the mouth of the Manning River between 1998 and 2001 (Hole *et al.* 2001) and Port Stephens (Murray 2014), as mentioned in this paper.

There have been few detailed studies on the diet of the beach stone-curlew, which is mainly understood from opportunistic observations. The records mentioned in this paper that contain some notes on feeding are almost entirely observations of beach stone-curlews hunting soldier crabs (*Mictyria* spp.) (Murray 2013; Birding-Aus 2016; Eremaea Birdlines 2016; D. Paton, pers. comm). Previous authors also recorded soldier crabs in the diet (Clancy 1986; Geering 1988; Woodall and Woodall 1989; Mellish and Rohweder 2012). The apparent significance of soldier crabs to the diet may be biased toward the increased visibility of this type of prey. Further studies are needed to confirm whether soldier crabs actually comprise the majority of the diet. Hole *et al.* (2001) introduced the possibility of beach stone-curlews raiding eggs of other ground-nesting birds, especially when crabs were few. The evidence for this was total breeding failure in little terns and red-capped plovers (*Charadrius ruficapillus*) coinciding with the period of time beach stone-curlews occurred nearby and beach stone-curlew tracks indicating their visitation of nests. This may account for observations at Lake Wollumboola and Burrill Lake of beach stone-curlews being mobbed by nesting birds (OEH 2015; Eremaea Birdlines 2016). Perhaps coincidentally, both these accounts occurred in January 2013.

Observations of breeding activity in Port Stephens (Murray 2014) are of great significance. Previously, the most southerly known breeding pair of beach stone-curlews was the pair at the mouth of the Manning River (Hole *et al.* 2001). The breeding pair at Port Stephens represents the only breeding individuals known in the Sydney Basin bioregion. The records presented in this paper were predominately sightings of single individuals. Excluding the birds at Port Stephens, pairs were seen in only two occurrence events. These were the beach stone-curlews at Toragy Point in December 2013 (Eremaea Birdlines 2016) and Merimbula in April 2016 (Birding-Aus 2016). The potential for the formation of breeding pairs in either account is not known.

The main threats to the beach stone-curlew include habitat destruction, low reproductive rate and increased predator populations (Marchant and Higgins 1993; Hole *et al.* 2001). Disturbance by beach users has also became a concern (NSW Scientific Committee 2008; Anon. 2016), however high visitation to beaches inhabited by beach stone-curlews may not necessarily have a significant impact, at least not in parts of Australia where the species is regularly reported (Freeman 2003). Clancy (1986) also noted that nests were being raided by egg-collectors. A proposed expansion of the marina at Soldiers Point in 2014 became contentious due to the residence of the beach stone-curlews at the site (Vernon 2014; HBOC 2015). The construction was proposed to take place on the beach where the birds regularly forage.

The regular observations on the beach stone-curlews at Port Stephens offers an excellent opportunity to study aspects of their ecology that are not well understood, such as fledgling diets, development of hunting behaviour, daily foraging patterns and seasonality of breeding events (Marchant and Higgins 1993; Mellish and Rohweder 2012). The patterns of records indicate that beach stone-curlew sightings in the Sydney Basin and South East Corner bioregions may increase in the future. Information on the ecology of the beach stone-curlew is therefore important to species management in this new extension of distribution.

ACKNOWLEDGEMENTS

I thank Alan Morris, Duade Paton, Barbara Jones, W. Ashley Hewson and John Bundock for useful discussions and the Cumberland Bird Observers Club and Far South Coast Birdwatchers Inc. for extracting sighting records. I also acknowledge all observers that have contributed to various databases. Elouise Sciacca, Peter and Antonia Hayler and Richele West are thanked for supporting this paper. Useful comments from two anonymous referees improved the manuscript.

REFERENCES

Amiet, L. (1957). A wader study of some Queensland coastal localities. *Emu* 57, 236-254.

Anonymous. (2016). 'Please give rare bird space.' Merimbula News. 21 March 2016. http://www.merimbulanewsweekly.com.au/story/3802895/please-give-rare-bird-space

Atlas of Living Australia (ALA). Database available at http://www.ala.org.au. Accessed 1 August 2016.

BEACH STONE-CURLEWS IN SOUTH EASTERN NEW SOUTH WALES

Birding-Aus. Database available at http://birding-aus.org. Accessed 1 August 2016.

Blakers, M., Davies, S.J.J.F. and Reilly, P.N. (1984). 'The Atlas of Australian Birds.' (Melbourne University Press: Melbourne).

Breen, D.A. (2007). Systematic conservation assessments for marine protected areas in New South Wales. Ph.D. Thesis, James Cook University, Townsville, Qld.

Clancy, G.P. (1986). Observations of nesting Beach Thick-knees *Burhinus neglectus* at Red Rock, New South Wales. *Corella* 10, 114-118.

Clancy, G.P. and Christiansen, M. (1980). A breeding record of the Beach Stone-curlew at Red Rock New South Wales. *Australian Birds* 15, 5.

Clarke, R. and Dolby, T. (2014). 'Finding Australian Birds: a Field Guide to Birding Locations.' (CSIRO Publishing: Collingwood, Vic.)

Cocker, M. (2013). President's Report. *Illawarra Birding* 18, 2.

Department of Environment, Climate Change and Water (DECCW). (2011). The Vertebrate Fauna of Royal and Heathcote National Parks and Garawarra State Conservation Area. NSW Department of Environment, Climate Change and Water, Hurstville, NSW.

Eco Logical Australia. (2011). Narrabeen Beach Dog Off-leash Exercise Area – Flora and Fauna Assessment. Report for Warringah Council. Eco Logical Australia, Sydney.

Eremaea Birdlines. Database available at http://www.eremaea.com. Accessed 1 August 2016.

Freeman, A.N.D. (2003). The distribution of Beach Stone-curlews and their response to disturbance on far north Queensland's Wet Tropical Coast. *Emu* 103, 369-372.

Garnett, S. (1993). Threatened and Extinct Birds of Australia. Report 82. Royal Australasian Ornithologists Union, Melbourne.

Garnett, S.T. (1992). Threatened and extinct birds of Australia. RAOU Report 82. Royal Australasian Ornithologists Union, Melbourne.

Garnett, S.T. and Crowley, G.M. (2000). 'The Action Plan for Australian Birds 2000.' (Environment Australia: Canberra.)

Geering, D.J. (1988). Some notes on a Beach Thick-knee runner. *Corella* 12, 22-24.

Hole, H., Hole, B. and Mardell, C. (2001). Observations of nesting Beach Stone-curlews on the Mid-north Coast of New South Wales, 1998-1999. *Australian Bird Watcher* 19, 49-54.

Hoskin E.S., Hindwood K.A. and McGill A.R. (1991). 'The Birds of Sydney.' (Surrey Beatty and Sons Pty Ltd: Chipping Norton, NSW.)

Hunter Birds Observers Club (HBOC). (2015). Submission on Proposed Soldiers Point Marina Extension – October 2015. Hunter Birds Observers Club, New Lambton, NSW.

Marchant, S. and Higgins, P.J. (Eds.) (1993). 'Handbook of Australian, New Zealand and Antarctic Birds.'

Volume 2. Raptors to Lapwings.' (Oxford University Press: Melbourne.)

Mayo, L.M. (1925). Large-billed Stone-curlew (*Orthorhaphus magnirostris*) in Moreton Bay. *Emu* 25, 40.

Mellish, G.F. and Rohweder, D.A. (2012). Reconstructing the diet of the Beach Stone-curlew *Esacus magnirostris* using scat analysis. *Australian Field Ornithology* 29, 201-209.

Morgan, J. (2013). What to expect for summer. *Eurobodalla Natural History Society Newsletter* 159, 3.

Morris, A., Charles, C. and Blanchflower, S. (2011). Unusual records June-Aug 2011. *Birding New South Wales Newsletter* 247, 15-18.

Morris, A.K., McGill, A.R. and Holmes, G. (1981). 'Handlist of the Birds of New South Wales.' (NSW Field Ornithologists Club: Sydney.)

Murray, J. and Dessmann, J. (2012). Flora and Fauna Assessment – Caltex Refineries (NSW) Pty Ltd – Kurnell Refinery Conversion. Report for URS Australia Pty Ltd. Biosis Pty Ltd, Sydney.

Murray, T. (2013). Soldiers Point Beach Stone-curlews. *Hunter Bird Observers Club Newsletter* 5, 13.

Murray, T. (2014). Soldiers Point Beach Stone-curlew update. *Hunter Bird Observers Club Newsletter* 2, 8-9.

NSW National Parks and Wildlife Service (NSW NPWS). (2003). The Bioregions of New South Wales: their Biodiversity, Conservation and History. NSW NPWS, Hurstville, NSW.

NSW Scientific Committee. (2008). Beach Stone-curlew *Esacus neglectus*. Review of current information in NSW. April 2008. Unpublished report. NSW Scientific Committee, Hurstville, NSW.

Office of Environment and Heritage (OEH). (2013). The Vertebrate Fauna of Towra Point Nature Reserve. NSW Office of Environment and Heritage, Sydney.

Office of Environment and Heritage (OEH). (2015). Pied Oystercatchers. *Shorebird Recovery Newsletter South Coast* 2014-2015, 11-16.

Office of Environment and Heritage (OEH). Atlas of NSW Wildlife available at http://www.bionet.nsw.gov.au. Accessed 1 August 2016.

Rohweder, D.A. (2003). A population census of Beach Stone-curlews *Esacus neglectus* in New South Wales. *Australian Field Ornithology* 20, 8-16.

Sapphire Coast Regional Science Hub (SCRSH). (2015). Beach Stone-curlew. Rare sight. http://www.atlasoflife.org.au/beach-stone-curlew-rare-sighting

Schulz, M. and Magarey, E. (2012). Vertebrate fauna: a survey of Australia's oldest national park and adjoining reserves. *Proceedings of the Linnean Society of New South Wales* 134, B215-B247.

Schulz, M. and Ransom, L. (2010). Rapid fauna habitat assessment of the Sydney metropolitan catchment area. In 'The Natural History of Sydney'. (Eds D. Lunney, P. Hutchings and D. Hochuli.) pp. 371-401. (Royal Zoological Society of New South Wales: Mosman, NSW.)

Smith, P. (1991). The Biology and Management of Waders
(Suborder Charadrii) in New South Wales. Species
management report 9. New South Wales National
Parks and Wildlife Service, Hurstville, NSW.

Stringfellow, D.S. (1962). Bar-shouldered Dove and
Frigate-birds in mid-coastal New South Wales. *Emu*
62, 65.

Trainor, C.R. (2005). Waterbirds and coastal seabirds of
Timor-Leste (East Timor): status and distribution
from surveys in August 2002–December 2004.
Forktail **21**, 61-78.

Vernon, J. (2014). Concerns over Port Stephens marina
expansion. ABC News. http://www.abc.net.au/
news/2014-03-29/concerns-over-port-stephens-
marina-expansion/5352974

Watkins, D. (1993). 'A National Action Plan for Shorebird
Conservation in Australia.' (Australasian Wader
Studies Group: Melbourne.)

Wilson, B. (1961). Beach Curlew in New South Wales.
Emu **61**, 64.

Woodall, P.F. and Woodall, L.B. (1989). Daily activity and
feeding behaviour of Beach Thick-knee *Burhinus
neglectus* on North Keppel Island. *Queensland
Naturalist* **29**, 71-75.